THE DANGER OF WORDS

STUDIES IN
PHILOSOPHICAL PSYCHOLOGY

Edited by
R. F. HOLLAND

THE DANGER OF WORDS

by
M. O'C. DRURY

LONDON
ROUTLEDGE & KEGAN PAUL
NEW YORK: HUMANITIES PRESS

266

First published in 1973
by Routledge & Kegan Paul Ltd
Broadway House, 68-74 Carter Lane,
London EC4V 5EL
Printed in Great Britain by
The Camelot Press Ltd, London and Southampton

© *M. O'C. Drury 1973*

ISBN 0 7100 7596 0

15
192
150.19

CONTENTS

PREFACE

The title of this book will at least indicate the hesitation I have long felt in putting forward these fragments for publication. They were written to be spoken aloud; hence the colloquial style, in many ways unsuitable for reading in the study. They were written for special occasions and with a specific audience in mind; hence the assumption of technical terms which are not otherwise defined. They were written to inaugurate a discussion; hence the incomplete manner in which every topic is left. For I have long held that discussion face to face is the proper medium for philosophy. Wittgenstein used to say that a philosopher who did not join in discussions was like a boxer who never went into the ring.

Why then do I now bring these papers together? For one reason only. The author of these writings was at one time a pupil of Ludwig Wittgenstein. Now it is well known that Wittgenstein encouraged his pupils (those at least whom he considered had no great originality in philosophical ability) to turn from academic philosophy to the active study and practice of some particular avocation. In my own case he urged me to turn to the study of medicine, not that

I should make no use of what he had taught me, but
rather that on no account should I 'give up thinking'.
I therefore hesitatingly put these essays forward as an
illustration of the influence that Wittgenstein had on
the thought of one who was confronted by problems
which had both an immediate practical difficulty to
contend with, as well as a deeper philosophical per-
plexity to ponder over. I do not of course claim
Wittgenstein's authority for a single idea expressed in
these papers. So far as I can remember I never discussed any of the topics touched on with him. They
were all written in the last few years, that is to say
well over a decade after his death. For what is written
here I take full responsibility, and write only what
seems to me to be the truth, and which I would be
prepared to defend in discussion. But that it was the
profound influence that Wittgenstein had on me as a
student that has developed into these reflections, of
that also I am certain. So perhaps then I can bring
some unity into what must otherwise appear very
fragmentary, if in this Preface I say something concerning the orientation that Wittgenstein gave to my
outlook.

For me from the very first, and ever since, and still
now, certain sentences from the *Tractatus Logico-
Philosophicus* stuck in my mind like arrows, and have
determined the direction of my thinking. They are
these:

> Everything that can be put into words can be put
> clearly.

> Philosophy will signify what cannot be said by pre-
> senting clearly what can be said.

> There are, indeed, things which cannot be put into
> words. They make themselves manifest. They are
> what is mystical.

This would not be the place, nor would I have the
ability, to discuss the differences and developments
which can be found between the *Tractatus Logico-
Philosophicus* and the *Philosophical Investigations*.
But this I must place on record. When Wittgenstein
was living in Dublin and I was seeing him constantly
he was at that time hard at work on the manuscript
of the *Investigations*. One day we discussed the
development of his thought and he said to me (I can
vouch for the accuracy of the words): 'My fundamental ideas came to me very early in life.' Now
among these fundamental ideas I would place the
sentences I quoted above. I think perhaps the remark
that Wittgenstein made, that after his conversations
with Sraffa he felt like a tree with all its branches
lopped off, has been misinterpreted. Wittgenstein
chose his metaphors with great care, and here he says
nothing about the roots or the main trunk of the tree,
these – his fundamental ideas – remain I believe
unchanged.

So now I want to say something about the word
'clarity' as it was understood by Wittgenstein. I owe
it to Mr Rush Rhees that he drew my attention to,
and translated for me, a remark that Wittgenstein
wrote in his notebook in 1930:

> Our civilisation is characterised by the word progress.
> Progress is its form: it is not one of its properties that
> it progresses. It is typical of it that it is building, con-
> structing. Its activity is one of constructing more

and more complex structures. And even clarity serves only this end, and is not sought on its own account. For me on the other hand clarity, lucidity, is the goal sought.

In this distinction between the two uses of clarity I see a difference of the very greatest importance. Let me make the point clear by a reminiscence. At one time for a short period Wittgenstein got me to read aloud to him the opening chapters of Frazer's *Golden Bough*. Frazer thinks he can make *clear* the origin of the rites and ceremonies he describes by regarding them as primitive and erroneous scientific beliefs. The words he uses are, 'We shall do well to look with leniency upon the errors as inevitable slips made in the search for truth.' Now Wittgenstein made it clear to me that on the contrary the people who practised these rites already possessed a considerable scientific achievement: agriculture, metalworking, building, etc., etc.; and the ceremonies existed alongside these sober techniques. They were not mistaken beliefs that produced the rites but the need to *express* something; the ceremonies were a form of language, a form of life. Thus today if we are introduced to someone we shake hands; if we enter a church we take off our hats and speak in a low voice; at Christmas perhaps we decorate a tree. These are expressions of friendliness, reverence, and of celebration. We do not believe that shaking hands has any mysterious efficacy, or that to keep one's hat on in church is dangerous!

Now this I regard as a good illustration of how I understand clarity as something to be desired as a goal, as distinct from clarity as something to serve a

further elaboration. For seeing these rites as a form
of language immediately puts an end to all the
elaborate theorising concerning 'primitive mentality'.
The clarity prevents a condescending misunderstanding, and puts a full-stop to a lot of idle speculation.

I would dwell a little longer on the distinction
between these two kinds of clarity, for it is a distinction that I hope will make my subsequent papers a
little clearer in their intention.

At one time I told Wittgenstein of an incident that
seemed to interest and please him. It was when I was
having my oral examination in physiology. The
examiner said to me: 'Sir Arthur Keith once remarked
to me that the reason why the spleen drained into the
portal system was of the greatest importance; but he
never told me what that importance was, now can
you tell me?' I had to confess that I couldn't see any
anatomical or physiological significance in this fact.
The examiner then went on to say: 'Do you think there
must be a significance, an explanation? As I see it
there are two sorts of people: one man sees a bird
sitting on a telegraph wire and says to himself: "Why
is that bird sitting just there?", the other man replies
"Damn it all, the bird has to sit somewhere".'

The reason why this story pleased Wittgenstein
was that it made clear the distinction between scientific clarity and philosophical clarity. Let me explain
this by an example of my own. Astronomers are
rightly interested in finding an explanation for the
remarkable 'red shift' in the spectral lines of the very
distant nebulae. The generally accepted explanation
is that it is a manifestation of the Doppler effect
(which is familiar enough to us in the way the pitch

of a rapidly moving whistle of a train changes as the train either approaches or recedes from us). So it is thought that these nebulae are receding from us at prodigious speeds. Now this is a possible scientific explanation, and in one sense it makes the phenomena clear. But then we at once want to ask, 'Why are all these nebulae receding at such speeds?' And this shows us that ultimately we will have to accept some facts as unexplained, and say, 'Well that is just how it is'. So then there would be nothing illogical in saying of the shift in the spectral lines, 'That is just how the spectra of distant nebulae are', and we are not *forced* to give any explanation. Philosophical clarity then arises when we see that behind every scientific construction there lies the inexplicable.

> The whole modern conception of the world is founded on the illusion that the so-called laws of nature are the explanation of natural phenomena.

Scientific explanations lead us on indefinitely from one inexplicable to another, so that the building grows and grows and grows, and we never find a real resting place. Philosophical clarity puts a full-stop to our enquiry and restlessness by showing that our quest is in one sense mistaken.

I would just give a few instances of reminiscences of conversations with Wittgenstein where a remark of his introduced sudden philosophical clarity by means of a full-stop.

> I told him I was reading a book about the 'Desert Fathers', those heroic ascetics of the Egyptian Thebaid. And, in the shallowness of those days, said

something to the effect that I thought they might have made better use of their lives. Wittgenstein turned on me angrily and said, 'That's just the sort of stupid remark an English parson would make; how can you know what their problems were in those days and what they had to do about them?'

He told me that he had just finished reading a book in which the author blamed Calvin for the rise of our present bourgeois capitalist culture. He said that he realised how attractive such a thesis could seem, but he for his part 'wouldn't dare to criticise a man such as Calvin was'.

Someone was inclined to defend Russell's writings on marriage, sex, and 'free love': Wittgenstein interposed by saying: 'If a person tells me he has been to the worst of places I have no right to judge him, but if he tells me it was his superior wisdom that enabled him to go there, then I know that he is a fraud.' He went on to say how absurd it was to deprive Russell of his Professorship on 'moral grounds'. 'If ever there was such a thing as an an-aphrodisiac it is Russell writing about sex!'

We had a discussion about the difficulty of reconciling the discourses and history in the fourth Gospel with the other three. Then he suddenly said: 'But if you can accept the miracle that God became man all these difficulties are as nothing, for then I couldn't possibly say what form the record of such an event would take.'

I have been trying to draw from my memories incidents which illustrate the conception of saying something clearly; bringing what looked like being a long and controversial discussion to a full-stop. It would be a contradiction to go on and say anything about

'There are indeed things which cannot be put into words'. But I would draw attention to this. In a letter written when he wanted to get the *Tractatus* published he said that 'it was really a book about ethics', and that the most important part is what is not said in it. I find myself thinking about this remark when reading all that he subsequently wrote. When he was hard at work on the manuscript of the *Philosophical Investigations* he said to me: 'I am not a religious man, but I can't help seeing every thing from a religious point of view.' And on another occasion: 'It is impossible for me to say one word in my book about all that music has meant in my life; how then can I possibly make myself understood?' And with regard to music this, which I have mentioned on another occasion: 'Bach put at the head of his *Orgelbüchlein*, "To the glory of the most high God, and that my neighbour may be benefited thereby". I would have liked to be able to say this of my work.'

I fear that these papers may be too metaphysical to be of interest to my colleagues occupied with the day-to-day problems of mental illness; and their topic too circumscribed and limited to interest philosophers. They are certainly not intended as in any sense a commentary on Wittgenstein's philosophy, but with the increasing importance that is now being given to his writings, they may possess a peripheral interest, as an illustration of his influence on one particular pupil. But then it must be added that all his life Wittgenstein was very dubious as to whether his influence on others (and on contemporary philosophy) was not more harmful than beneficial.

ACKNOWLEDGMENTS

The translations of Lichtenberg's Aphorisms which appear in this book are taken from Dr J. P. Stern's *Lichtenberg, a Doctrine of Scattered Occasions*. I wish to thank Dr Stern and his publishers, Thames and Hudson, and Indiana University Press, for permission to use these translations.

In the chapter on 'Hypotheses and Philosophy' the material for criticising the mutation-selection theory of evolution is deeply indebted to Professor C. P. Martin's important book entitled *Psychology, Evolution and Sex*. I am grateful to Professor Martin and to his publisher, Charles C. Thomas, Springfield, Illinois, for allowing me to use this material and to quote from the book itself. I would like to thank Professor J. C. Eccles and his publishers, the Clarendon Press, for permission to quote from his book *The Neuro-Physiological Basis of Mind*, and Gallimard for permission to quote Simone Weil's 'Lettre à une élève' from *La Condition Ouvrière*.

In conclusion I would like to express my gratitude to Professor R. F. Holland for all his valuable help in preparing my manuscript for publication.

Nous savons au moyen de l'intelligence que ce que l'intelligence n'appréhende pas est plus réel que ce qu'elle appréhende.

SIMONE WEIL

CHAPTER 1

WORDS AND TRANSGRESSIONS

> I WOULD counsel all young people to put all new
> words in careful order and arrange them like
> minerals, in their various class, so that they can be
> found when asked for or when required for one's own
> use. This is called word economy, and is as lucrative
> to the mind as money economy is to the purse.
>
> LICHTENBERG

In the Proverbs 10:19 it is written: 'With a multitude
of words transgressions are increased.' And I will
make this text an excuse for the substance of this
paper. For I want to speak to you about the way
words can lead us into confusion, misunderstandings,
error. Confusion when we are talking to patients,
misunderstandings when we discuss mutual problems
with our colleagues, error when in solitude we try to
clarify our own thinking.

For the purpose of classification I have divided these
fallacies under five separate headings. The first I call
the fallacy of the alchemists; the second the fallacy
of Molière's physician; the third the fallacy of Van
Helmont's tree; and the fourth the fallacy of the missing hippopotamus; finally, for the fifth, I have chosen
the unoriginal title of the fallacy of Pickwickian senses.

BDW

First then the fallacy of the alchemists. I have chosen this name because of what Lavoisier says in the Introduction to his *Treatise on Chemistry*. Lavoisier, you will remember, was the first chemist to introduce our modern system of nomenclature into chemistry; a system whereby different substances are named in terms of the elements which go to form them. Sodium chloride, potassium permanganate, calcium carbonate, etc., etc. Prior to this book many of these substances, though well known, had bizarre names which in no way indicated their relationship to each other. These names often go back to the days of the alchemists; sometimes indicative of their original discoverer, sometimes from their place of origin, sometimes to some irrelevant outward appearance. Thus we had 'Glauber's salts', 'Fuming liquor of Libavius', 'Butter of arsenic', 'Vitriol of Venus', and so on. In his Introduction Lavoisier says that he is now introducing a *method of naming* as distinct from a *nomenclature*, and he adds the following wise remark about the importance of what he is now doing:

> If languages really are instruments fashioned by men to make thinking easier, they should be of the best possible kind, and to strive to perfect them is indeed to work for the advancement of science. For those who are beginning the study of a science the perfecting of its language is of high importance.

And later on he writes:

> It is therefore not surprising that in the early childhood of chemistry, suppositions instead of conclussions were drawn; that these suppositions transmitted from age to age were changed into presumptions, and

that these presumptions were then regarded as fundamental truths by even the ablest minds.

Now I think if we are to be honest with ourselves
we must admit that the vocabulary of psychiatry
today is only too comparable with what Lavoisier has
to say about the nomenclature of chemistry in its
childhood. We have indeed a *nomenclature*, but we
have no *system of naming*. Some diseases are named
after those famous physicians who first described
them; thus we have Korsakov's psychosis, Alzheimer's
disease, Ganser's syndrome; some are named in terms
of a long discarded pathology, hysteria for example
and schizophrenia. Some are named in terms of the
most prominent symptom; the word 'depression' is
used both for a complaint of the patient and for the
diagnosis of the attending physician; similarly with
the words 'anxiety state'.

I would have to agree that having no better terminology at hand we must for the present do the best with
what we have. But let us be on our guard against those
dangers that Lavoisier warned us against. Let us
beware lest from this unsystematic nomenclature
suppositions are drawn, which then become presumptions and only too easily pass over into established truths.

I would say that the chief danger of an unsystematic
nomenclature is the danger of regarding its classifica-
tions as mutually exclusive and completely exhaustive.
For example it is only too easy to get involved in a
controversy as to whether this patient is a schizophrenic or a case of endogenous depression, when for
all we know he might be both at the same time; or

neither, but some other disease for which we have at present no convenient name.

There is a story told of a candidate up for the membership examination who in answer to one of the more difficult questions on the paper could only reply indignantly, 'This is not mentioned in Tidy's *Synopsis of Medicine*.' It is important for us to bear in mind that there are still many diseases both of mind and body which are not only not mentioned in Tidy's *Synopsis*, but are not in *any* text-book or encyclopaedia of medicine. The science of medicine, and particularly psychiatry, is not yet complete.

Janet wrote an interesting essay on the history of the word 'neurosis'. He showed that in spite of various attempts to define the limits of this term, in practice the word has been used to cover all those clinical conditions which at the time of writing could not be accounted for by any known pathology. Thus the famous Pinel, in whose days the ophthalmoscope had not been invented, classified all cases of blindness in which there was no manifest disease of the external eye, cornea, or lens, as hysterical amaurosis. Trousseau, that prince of clinicians, after giving a masterly description of the symptoms and signs of *tabes dorsalis*, classifies it as a form of neurosis. For in his day there was no known method of staining nervous tissue and demonstrating the degeneration of the posterior columns of the spinal cord. Other equally able writers have in their time classified Parkinson's disease, Grave's disease, hydrophobia, tetanus, eclampsia, as psychogenic in origin.

Now these things were written for our learning. We are certainly making similar mistakes today. Considering the immense complexity of the anatomy, physiology, and bio-chemistry of the human body, it is certain that there are probably more diseases still to be described than have so far been given a precise description and a name. For example the estimation of the blood sugar is a comparatively recent achievement. In the past, spontaneous hypo-glycaemia certainly occurred, and the mental and behavioural disorder so produced was described as neurotic, psychotic or epileptic, according to the degree and rapidity of the hypo-glycaemia involved. Wisdom demands that we remember constantly our ignorance.

This then brings me on to the second danger of words, that which I have called the fallacy of Molière's physician.

In one of his plays Molière has a physician asked this question: 'How is it that opium is able to put people to sleep?' The physician replies with great profundity that it is because opium has 'dormitive properties', and this answer is found entirely satisfactory by his interlocutors. I think we all have a tendency to deceive ourselves in this way. To use obscure and learned phrases, thinking thereby that we have obtained a deeper insight. I remember as a medical student reading the chapter on fractures in a manual of surgery; it began by stating that by a fracture is meant 'the dissolution of continuity in a bone'. This struck me as quite as funny as Molière's joke. It is a wise rule from time to time to force oneself to write down in simple language the precise meaning of any involved circumlocution we have got into the habit of using. If we did this we would find, I think, that such words as 'hysteria', 'psychopathic personality', 'character neurosis', are symbols of our ignorance rather than of any understanding.

Let me for a moment give you an example of a more serious error of this type. It was for a time fashionable, and still is, I am told, to produce a dramatic emotional reaction in a patient by getting him or her to inhale a mixture of carbon dioxide and oxygen; and other chemical means either by inhalation or injection were also used. Such a reaction was given the profound sounding name of 'abreaction', and this technical term led people to believe that they understood what was happening; that the patient undergoing this treatment was releasing forces and tendencies which in his previous state were repressed and causing his symptoms, and therefore that such abreaction was certain to be both informative to the psychiatrist and beneficial to the patient. One of my colleagues told me that he saw this treatment being administered to a rather timid little man who was a victim of alcoholism. When the mask through which the carbon dioxide mixture was administered was firmly held over his face, this little fellow fought back with surprising fury. Ah! said the psychiatrist, now you see that the cause of his addiction is this deeply repressed aggressiveness. On hearing this story I was reminded of Voltaire's remark: 'This animal is very dangerous, when it is attacked it defends itself.'

You might reply to me that unless we experiment in this way research in psychiatry and in medicine generally will come to a standstill. And so I must now speak about that most dangerous word in our present medical vocabulary, 'research'. I am informed that something in the region of a million new scientific papers are published in the journals every year, and that these if they were all to be bound in one volume would be equivalent to three complete editions of the *Encyclopaedia Britannica*. Gentlemen, I do not believe we are living in an age of such colossal originality. Let this be clearly said: research in the proper meaning of the much abused word does not mean *collecting facts*; there is much too much fact collecting going on. Research means new ideas; new concepts, new ways of looking at old and familiar facts. The important part of research is the thinking done *before* the experimental verification gets under way.

The ability to think in this particular way is, I believe, a comparatively rare talent. A gift for research is not the automatic accompaniment of a grant for research. There is, I suppose, no more honoured name in the history of physiology than that of Claude Bernard. At one time when a prolonged illness prevented him continuing his experiments in the laboratory, Bernard composed a short treatise setting out the principles which had guided him in making his discoveries. This volume, though now over a hundred years old, contains much that needs repeating today. If I may just quote a few short passages from it you will perceive how apposite it is.

Bernard writes:

Two operations must therefore be considered in any experiment. The first consists in *premeditating* and bringing to pass the conditions of the experiment; the second consists in noting the results of the experiment. It is impossible to devise an experiment without a preconceived idea; devising an experiment, we said, is putting a question; we never conceive a question without an idea which invites an answer. I consider it therefore an absolute principle that experiments must always be devised in view of a preconceived idea, no matter if the idea be not very clear or well defined. As for noting the results of the experiment, which is itself only an induced observation, I posit it similarly as a principle that we must here, as always *observe* without a preconceived idea.

Once in conversation with a friend Bernard put the same important principle in a more aphoristic form. 'When you go into the laboratory do not forget to leave your imagination in the ante-room with your overcoat; on the other hand never forget to take it away with you when you go home.'

There are two further points I would just touch on concerning which Bernard's teaching is much needed as a corrective to what often passes under the name of research today.

He writes:

Misconceived erudition has been, and still is, one of the greatest *obstacles* to the advancement of experimental science.

Now if you pick up any modern scientific journal it seems almost a standard practice for the author to start with a review of the previous literature. Thus often such articles have a hundred or more references at the end, and it is even worse when it comes to the bibliography at the end of some books. I am inclined when I see such a display of erudition to pass on to something more profitable. For I fear that such an author will have his mind so constipated with facts as to be incapable of producing anything but wind. If an author really has a contribution of value to make, then let him get on with it at once, there is no need for him to give a display of his homework.

The second point that Bernard emphasises is in my opinion even more important for what goes by the name of research in the behavioural sciences today. He writes:

In every science we must recognise two classes of phenomena, those whose cause is already defined; next those whose cause is still undefined. With phenomena whose cause is defined statistics have nothing to do; they would even be absurd. As soon as the circumstances of an experiment are well known we stop gathering statistics . . . Only when a phenomenon includes conditions as yet undefined, can we compile statistics; we must learn therefore that we compile statistics *only when we cannot possibly help it*; for in my opinion statistics can never yield scientific truth, and therefore cannot establish any final scientific method.

Statistics can bring to birth only conjectural sciences; they can never produce active experimental sciences, i.e. sciences which regulate phenomena according to definite laws. By statistics we get a conjecture of greater or less probability about a given case, but never any certainty, never any absolute determinism. Of course statistics may guide a physician's prognosis; to that extent they may be useful. I do not therefore reject the use of statistics in medicine, but I condemn *not trying to get beyond them* and believing in statistics as the foundation of medical science.

I do not think that those wise words need any further

comment from me. But perhaps you would bear them in mind next time you find yet one more mass of statistical information in the *British Journal of Psychiatry*.

This gives me the cue to introduce the next fallacy I mentioned, the one I called the fallacy of Van Helmont's tree.

Van Helmont, as you know, was one of the great founders of chemistry. He was the first chemist to realise the importance of the chemical balance; of carefully weighing everything both before and after a chemical reaction. Indeed, it was largely due to his work that the principle of the conservation of matter became an established axiom. Now Van Helmont performed a certain experiment with great care and accuracy, whose result seemed irrefutable and yet at the same time absurd. It was this.

He weighed accurately a certain quantity of earth and placing it in a large pot, planted a small ash sapling. Every day he watered the plant with pure distilled water, and in between these waterings he kept the surface of the soil covered so that no foreign extraneous matter should fall on it. In due time the sapling grew to such a size that its weight had increased more than a hundredfold, in fact it had become too big for the pot to hold it. Van Helmont weighed it carefully, and then weighed the original soil he had filled the pot with, finding that this latter had lost nothing. He argued therefore that as the only additions made were those of pure water all the materials in the tree, bark, pith, leaves, etc., were in some way composed of nothing but water. This certainly seemed paradoxical both to him and his contemporaries, but the evidence of the experiment seemed irrefutable. Where did they go wrong? Well of course they did not know that a plant is able to extract carbon from the carbon dioxide of the air by the process of photosynthesis; the very existence of such a substance as carbon dioxide or such a process as photosynthesis was then undreamed of. Similarly how could they have guessed that there were minute organisms in the soil that could extract nitrogen from the air and transmit it to the plant?

Now the motto of this is that in the early stages of any science when there are still a host of unknown factors at work it can be most misleading to draw conclusions from experiments however accurately performed. The methods employed may be too precise for the data on which they have to work. I am told that today if you wish to get any report on the use of a new method of treatment it is essential that the investigation be carried out on the basis of what is known as a 'double blind trial'. When I hear this I murmur to myself, 'Remember Van Helmont's tree.' For it seems clear to me that psychiatry is still dealing with too many unknowns for the method of the double blind trial to be either safe or applicable. I speak here only of the logic involved, and say nothing of the ethical aspect of doctors deliberately allowing themselves to be in ignorance of the treatment their patients are receiving.

The logical essential for the double blind trial to be in any way convincing is that the experimental group and the control group should be evenly matched. This today means matched as to age, sex, duration of illness, nature of symptoms, previous treatments given. But it may well be that these are not the necessary factors alone. May it not be that there are a host of genetic, bio-chemical, histological factors that also need to be taken into account and of which we are at present totally unaware?

Perhaps I can make this point clearer by an imaginary example. In the seventeenth century 'being sick of a fever' was a respectable diagnosis for a physician to make. It was not known nor even guessed that the important factor was the micro-organism causing the fever. Quinine had recently been introduced into Europe, and malaria being more common in these parts than it is now, quinine soon proved its usefulness. Now suppose some physicians had in those days said: we must now have a double blind trial to make sure that quinine is not just a placebo. Those whose trial contained many cases of malaria would have statistical proof of its efficiency. Whilst in another group there might be more cases of relapsing fever on which quinine would show no therapeutic benefit. Thus we would get two properly carried out double blind trials and contradictory results. This seems to me to be happening in the trial of new psycho-tropic drugs today. Statistically adequate trials, but contradictory results. Our psycho-pathology is not yet adequate to make such elaborate experiments justified. For as Osler said long ago: 'As is our pathology, so is our therapeutics.' The example I have given is of course only an imaginary one. So let me further emphasise the possibility of such fallacious argument from apparently irrefutable data by a real example from the history of psychiatry. That disease which we now call G.P.I. was first described and clearly differentiated as a clinical entity by French clinicians in the decade 1820 to 1830. Nearly all the cases first described were old soldiers from Napoleon's Grand Army. We know only too well why *that* should have been the case. But to those clinicians it seemed statistically self-evident that the cause of G.P.I. was the undermining of the nervous constitution of those who had endured the privations and the horrors of the retreat from Moscow. They knew nothing of a minute spirochaete; but the battle of Borodino and the crossing of the Beresina were still vivid memories.

So I would sum up the fallacy which I have entitled 'the fallacy of Van Helmont's tree' in these terms. Carefully planned and well executed investigations in the early stages of a science may be completely misleading, just because of our ignorance of the possible factors involved.

But what then are we to do? We are daily inundated with information concerning new drugs by those who with the best intentions are yet financially interested in their sale. Surely we must adopt some scientific procedure to sift out the good from the mediocre or even the useless. Yes, indeed, it is necessary that this should be done. But I do not believe the double blind trial is in the present state of our knowledge either scientific or helpful. During the twenty years or so that I have been working in psychiatry I have seen and used many forms of treatment; some I still use, others I have almost forgotten about until looking through some old notes I am reminded that for a short period they seemed to be worth trying. (For instance, a few days ago I had occasion to look up the notes of a patient whom I had not seen for many years, and I saw that when I last treated her I used a drug called 'Cavo-dil'; I had to think twice before I remembered that it was one of the M.A.O. group that enjoyed consider-able popularity for a year or so.) My experience has been that there is a process of *natural selection* at work in all forms of treatment, and that there is the survival of the fittest. And that this process of selection takes place without any one person or any particular investigation deciding the matter once and for all. Let me give you an example. When I came into psychiatry 'insulin coma therapy' was the best and only treat-ment we had for schizophrenia. Having worked for three years in an insulin coma clinic I am certain that it was worth doing, that it was better than doing nothing. With the introduction of chlorpromazine the number of cases coming for insulin coma gradually decreased until it became certain that we could dis-pense with insulin coma altogether. Now this was not *one* person's decision, nor was it one deliberately made on a *particular* occasion; it just happened. If we as clinicians continue to do our work with attention, with courage when it is needed, and with the necessary amount of scepticism, then this natural selection will continue to work for us. I am impressed when I read the works of some of the great clinicians of the last century, by their style of description. They do not hesitate to use such expressions as 'my experience has been', 'I have often noticed', 'the following case impressed me', etc. This would now be rejected by some as 'merely anecdotal evidence'. At the risk of appearing very old-fashioned I am going to claim that this keen attention to *anecdotes* is of the first impor-tance. I do not of course mean that we should publish these anecdotes; that would merely add to the confusion. But we should have our eyes and ears open, and our pens ready to note down in our case-books, every incident or remark that seems in any way novel or strikes our attention. I know for myself the danger of my case history taking to become stereotyped. I wish we all had more time to listen. Again if I can refer back to Claude Bernard's account of his own discoveries, he describes in detail many particular and unsought-for observations, which after reflection and speculation led him to a new hypothesis to be tested by a planned and crucial experiment. One of his pupils describes how during an experiment Claude Bernard seemed to have eyes all round his head, he would point out quite evident phenomena which no one else had noticed. Of course such a great hypothesis as 'the preservation of the constancy of the internal milieu' was the result of imagination and not of any one particular observation. But it was the ability by means of which this imagination was stirred into speculation by some one particular observation, that constituted him the great scientist that he was. Such minds are rare; it is probable that there are few in any one generation, but then the real contributions to the permanent advancement of science are equally rare.

I sometimes wish it was a law that every scientific paper had to be allowed to mature for ten years in bond, like good whisky, before being allowed in print.

These reflections bring me directly to a consideration of the fourth type of verbal fallacy, the one I have called 'the fallacy of the missing hippopotamus'.

I chose this rather bizarre name from a discussion which once took place between the two philosophers, Bertrand Russell and Ludwig Wittgenstein. Wittgenstein illustrated the point he wished to make by the following example. Suppose I state 'There is an hippopotamus in this room at this minute, but no one can see it, no one can hear it, no one can smell it, no one can touch it; have I now with all these added provisos said anything meaningful at all?' Surely not, for a proposition that can neither be verified nor refuted has no useful place in scientific language.

But I think you would be surprised to find how easy it is to make just this sort of logical error. Modern science is full of missing hippopotami. We are inclined to fall in love with an hypothesis, and so when facts begin to tell against it, we invent a subsidiary hypothesis to save the face of the first, and this process continues until without realising it our first hypothesis has become so secure as to be irrefutable. But alas, in doing just this we have at the same time deprived it of all significance.

Two examples from our own sphere of science will make my point clearer. Freud had the original and suggestive idea that dreams were really wish fulfilments, and not only that but always sexual wish fulfilments. Some dreams obviously are. But others on the face of it were not. So in order to save his beloved hypothesis he had to invent a great many subsidiary hypotheses, those that he described under the name of the dream mechanisms: condensation, displacement of affect, symbolism, etc., etc. He does not seem to me to have observed that in so introducing all these extra hypotheses he has emasculated his original idea of all significance. Let me make this clearer by an incident which Janet relates. Janet was talking to an enthusiastic pupil of Freud: 'Last night,' said Janet, 'I dreamt that I was standing on a railway station: surely that has no sexual significance.' 'Oh! indeed it has,' said the Freudian; 'a railway station is a place where trains go to and fro, to and fro, and all to and fro movements are highly suggestive. And what about a railway signal; it can be either up or down, need I say more?' Now as Janet rightly went on to point out, if you allow yourself such a freedom in symbolism, every possible content of any dream whatsoever can be forced into this type of interpretation. The theory has become 'fact proof'; it just can't be refuted. But that which cannot be proved wrong by any conceivable experience is without meaning. The object of a statement, of an hypothesis, is to state which of two possible alternatives is in fact the case. If *no* alternative is allowed, the statement decides nothing about a possible state of affairs.

One further example of the same verbal fallacy, one more missing hippopotamus. Wolpe put forward the interesting hypothesis that all neurotic behaviour was learnt behaviour, and that therefore such maladaptive behaviour could be 'cured' by the application of what certain psychologists have called 'learning theory'. Now for Wolpe and those who have taken up behaviour therapy with enthusiasm, learning is always a matter of establishing a stimulus-response connection. Therefore the first task of the behaviour therapist must be to ascertain the stimuli which have become unnecessarily linked with anxiety. But then,

CDW

as Wolpe has to admit, this is not always easy. It
is easy enough with a straightforward mono-
symptomatic phobia such for example as a phobia for
cats. Such mono-symptomatic phobias are not common, and when they do occur, such an aetiology and
therapy as Wolpe suggests is probably correct. But
what about that much more common syndrome which
Freud described under the very suitable name of
'free floating anxiety'? Here the specific stimulus
linked with the anxiety reaction is hard to discern.
But Wolpe has become too attached to his all-
embracing explanation of anxiety states. Free floating
anxiety, he claims, is anxiety linked with such ever
present stimuli as Space, Time, and the idea of Self!
Now if the word 'stimulus' is to be applied to such
concepts as these then the whole stimulus-response
theory of learning becomes irrefutable and at the same
time meaningless. The theory, as is well known,
depends on the experimental work with animals carried out originally by Pavlov, Thorndike, Hull,
Skinner, and their followers. One only has to ask
oneself to imagine these experimenters using Space,
Time, or the idea of Self, as a stimulus in any of their
experiments to see the enormous extrapolation that
Wolpe has here made. It is not possible to refute him
because he has said nothing. I cannot refrain from
quoting Janet once again. Janet built up a most
interesting psychology based on the twin concepts of
'psychic energy' and 'psychic tension'. In the Introduction to one of his books he makes the profound
remark that one great advantage of his theory is that
time may prove him to be completely wrong. As a
matter of fact I think Janet's hypothesis has not been

substantiated, and if he was still alive he would perhaps have replied, 'Well, what did I tell you'. I think we must all be on the watch that in psychology and psychiatry we take care to formulate hypotheses which are capable of being refuted. No more missing hippopotami please.

I come then, finally, to that fallacy which has now the name of the fallacy of 'Pickwickian senses'. The name was taken from a famous scene which took place one evening at the Pickwick Club. Mr Blotton had the termerity to call Mr Pickwick a humbug. This was the occasion for some heated words between various members, and order was only restored to the meeting when the chairman suggested to Mr Blotton that he had only used the term 'humbug' in a purely *Pickwickian* sense, and not with its usual connotation. Mr Blotton agreed that he had the highest regard for the honourable member Mr Pickwick and only described him as a humbug in a purely Pickwickian sense. After this explanation Mr Pickwick said he was completely satisfied with his friend's explanation and that he had used certain terms of abuse during the incident in a purely Pickwickian sense also. Peace was restored once more to the meeting.

Now I think that in psychiatry today we are inclined to use certain words in 'a purely Pickwickian sense' – words which to our patients sound as a reflection on their personality although we mean no such moral criticism. I have for many years tried to get the word 'alcoholic' dropped. For if we ask a patient to accept this description of himself he thinks of the familiar drunkard of literature and stage. One who is never quite sober, smelling of drink, a large red nose and blood-shot eyes, etc., etc. Now of course we don't mean this description at all when we diagnose the disease alcoholism. We know that a patient may be addicted to alcohol without ever having been intoxicated, who has had long periods of contented sobriety, who does not experience a constant craving for alcohol, who on examination may show no outward and visible signs of his illness. We mean by an alcoholic one whose pattern of drinking has developed certain sinister signs with which you are all familiar and which I therefore need not elaborate here. But to the patient we are using a term which reflects on his personal integrity. He or she does not realise that psychiatrists today use the term 'alcoholic' in a purely Pickwickian sense. We mean a person who either for metabolic or temperamental reasons (it is not yet known exactly which; or possibly both of these explanations apply) should be advised by his doctor to abstain from alcohol entirely. Surely it would be possible for us to find a name for this medical condition which would obviate so many unnecessary arguments. My own experience has been, when I tell a patient that alcoholism is not a scientific term, that therefore I am not calling him an alcoholic but I am advising him that he has shown signs which clearly prognosticate the need for total abstinence in the future, that such an explanation is more readily agreed to, or at least open to an intelligent discussion.

A similar state of affairs exists, I believe, with regard to the use of the word 'hysteria'. If we tell a patient that her symptoms are hysterical in origin, or even if we use this term in writing to her general practitioner, we will be taken to mean that the patient's condition need not be taken seriously, and that a dose of cold water either literally or metaphorically is all that is required. It is a word that should be dropped from a scientific vocabulary. I know that the word is often used by competent psychiatrists in a purely Pickwickian sense and without meaning to minimise the need for help and therapy. But I am also well aware in my own case of a strong temptation to label as hysterical all those symptoms for which I can find no good cause and where my therapeutic attempts have not met with success. It has been suggested, I know, that the word 'functional' should be used instead of 'hysterical'. This avoids the fallacy of 'Pickwickian senses' in that it will not offend the patient; but it is an example of the fallacy of Molière's physician in that it pretends to explain by a learned circumlocution a condition which to date neither doctors nor patients understand. I am not convinced that in psychiatry an air of omniscience and omnipotence is appreciated by the patient; more often all that is required is a concerned listening and an obvious attempt to do something helpful.

I am not sure, but I feel that the word 'depression' is beginning to be used in a Pickwickian sense, a sense in which the psychiatrist means one thing and the patient understands another. The development of effective treatment for the old-fashioned melancholia or manic-depressive psychosis has contributed to this state of affairs. For we now find that these treatments are effective in some conditions where 'depression' or 'melancholy' are not complained of by the patient. Hence if we tell him that he is 'depressed' he may well come to the conclusion that we are confused about his condition. There is a danger too that if we begin to talk about 'atypical depression', or 'masked depression', we may be committing that fallacy which I called the fallacy of the alchemists. There is a danger that this nomenclature may lead us into accepting as an established truth what at best is only a conjecture. In our present state of knowlege two conjectures are possible. One is that the illnesses which respond to the same treatment are all manifestations of one and the same underlying pathology. The other is that the new forms of chemotherapy are potent against a variety of different diseases. Penicillin can cure both a carbuncle and lobar pneumonia. It may be that the mental conditions which respond to the tri-cyclic thymoleptics are equally disparate. It is important that for future understanding we keep the choice between these two alternatives open.

One word more about the confusion that the word 'depression' can cause. Patients cannot be expected to understand that by the word 'depression' the psychiatrist understands a very different condition from that denoted by the word 'unhappiness'. It will inevitably happen from time to time that we will be asked and expected to remedy the normal discontents and disappointments that are part of our common human life. Freud showed real profundity when he stated that the aim of psycho-analysis was to replace neurotic unhappiness by normal unhappiness. A psychiatry based on a purely hedonistic ethics, a psychiatry that does not recognise that periods of anxiety and periods of melancholy are a necessary part of every human life, such a psychiatry will never be more than a superficial affair. Our task must be not only to relieve but also to interpret.

Another source of considerable confusion between doctors and patients today is in the use of the word 'drugs'. In strict etymology the word 'drug' means any measured quantity of medicine. So that a patient who is taking iron for anaemia, Vitamin B for neuritis, or insulin for diabetes, is receiving drug treatment. But of course in the popular mind the word 'drugs' has many frightening associations: drug addiction, being under the influence of drugs, the drug traffic, etc. In the popular mind a drug is something that is taken for its immediate soporific or stimulating effect. Now it is one of the happier features of this present age that chemical substances have been discovered which have a profound psycho-tropic effect. Both 'endogenous depression' and 'schizophrenia' can be treated by the prolonged administration of certain 'drugs'. But we must explain to our patients that these substances are not given for purely immediate sedative or stimulating effects; they are given over a prolonged time and only show their benefits after several weeks. My own belief is that these substances are more in the nature of replacement therapies, like the iron, the vitamin, the insulin I mentioned above. Hence when we tell a patient that we propose to treat him with drug therapy we must beware of the fallacy of 'Pickwickian senses', and explain to him the difference between a necessary chemical and a temporary anodyne. There is much talk nowadays about the spread of drug addiction, and I have known some patients who have given up their necessary medication because of what they have read in the popular press.

Of course it is best of all when a patient no longer needs a doctor or his prescriptions, but this is not always possible to achieve.

It is written in the Book of Proverbs that 'with a multitude of words transgressions are increased'. What an excellent motto that would be for our new Royal College of Psychiatry.

CHAPTER 2

SCIENCE AND PSYCHOLOGY

PEOPLE do not readily give up false opinions about man once they feel justified in claiming that they derive them from a subtle knowledge of humanity, and that only certain initiates are capable of such insights into the hearts of their fellow men. Consequently there are few branches of human knowledge where a little learning is more harmful than in this.

LICHTENBERG

Ladies and gentlemen,

I think I can best introduce the subject of my paper tonight by a series of quotations. Quotations taken from the writings of psychologists, who either in their own day, or at the present time, were and are recognised as authorities in the faculty of psychology.

My first quotation is from the justly famous American psychologist William James; the author of the *Principles of Psychology*, a book which is still well worth reading. But my quotation is this:

Psychology is not yet a science but only the hope of a science.

That was written in 1890. Thirty years later the

eminent French psychologist, Pierre Janet, concluded his two large volumes on psychological healing with the remark:

> Medical practitioners have suddenly turned to psychology, and have demanded of this science a service which the psychologists were far from being prepared to render. Psychology has not proved equal to the occasion and the failure of the science has thrown discredit upon psychotherapy itself. But this very failure has necessitated entirely new psychological studies, whereby the science of psychology has been regenerated. . . . Some day we may hope that there will be enough knowledge to make it possible to budget the income and expenditure of a mind, just as today we budget the income and expenditure of a commercial concern.

Thirty years later again Hebb published his book called *The Organisation of Behaviour*. This book had no small influence on future psychological thinking, and is still often quoted in the literature. In the Introduction to his book Hebb commences with this statement.

> It might be argued that the task of the psychologist, the task of understanding behaviour and reducing the vagaries of human thought to a mechanical process of cause and effect, is a more difficult one than that of any other scientist. Certainly the problem is enormously complex; and although it could be argued that the progress made by psychology in the century following the death of James Mill, with his crude theory of association, is an achievement scarcely less than that of the physical sciences in the same period, it is never the less true that psychological

theory is still in its infancy. There is a long way to go before we can speak of understanding the principles of behaviour to the degree that we understand the principles of chemical reaction.

Ten years later O. L. Zangwill in an article on psychology written for the 1950 edition of *Chambers' Encyclopaedia* stated:

At the present time it must be admitted that psychology falls sadly short of its aim. In view of the complexity of its data and the difficulties confronting crucial experiments in the psychological sphere, the explanations offered by mental science remain at the descriptive level. Hypotheses intending to co-ordinate large bodies of fact, such as the psycho-analytic or the gestalt theory, fall short of the necessary requirements for truly scientific precision. But the prevailing uncertainty of psychological explanation need imply no fault more severe than scientific immaturity. Indeed the contemporary situation in psychology is strikingly parallel to that of physiology in the sixteenth century. The notable development of modern experimental physiology leads one confidently to expect that a coherent science of mind will slowly take shape in the general framework of the sciences of life.

Ten years later again H. J. Eysenck in the Introduction to his big book on *Abnormal Psychology* writes as follows:

Originally I conceived the writing of a book such as this fifteen years ago when the exigencies of war threw me into contact with psychoneurotic patients at the Mill Hill Emergency hospital. Having little knowledge of the field, I naturally turned to the textbooks available on psychiatry and abnormal clinical psychology. The perusal of some fifty of these left me in a state of profound depression, as none of them contained any evidence of properly planned or executed experimental investigations, or even the realisation of the necessity for such. Nor did I find that concise and consistent framework of theories and hypotheses which usually precede experimental investigations; all was speculation and surmise, laced with reference to clinical experience. Michael Faraday's words seemed only too apposite: 'They reason theoretically without demonstrating experimentally, and errors are the result'. . . . It is for this reason that the dedication is to E. Kraepelin, the first person to be trained in the psychological laboratory, and to apply experimental methods to abnormal psychology. It is sobering to consider, if only his outlook had prevailed in psychiatry how much further advanced our knowledge would now be.

You will see, I think, from these quotations that there has been over the last eighty years or so a general agreement among psychologists that the really important work, the really significant discoveries, in the science of psychology, belong to the future. Psychology, they seem to agree, is still a very young science, but one that once it adopts the rigour of experimental science will bear great fruit. When Galileo performed his first experiments in rolling marbles down an inclined plane, and measured mass, time and velocity, he made the remark: 'This is the beginning of a great science.' And so indeed it was. It was the beginning of physics. And I need not remind you how today the science of physics has changed the whole manner of human life, and of our way of thinking about the nature of the universe we live in.

Unless I misunderstand them, the psychologists I have just quoted are encouraged by the hope that their rudimentary experiments – dogs salivating at the sound of a bell, rats learning to run a maze, pigeons learning to do strange tricks in Professor Skinner's box, human beings day-dreaming over ink-blots – that these experiments are the harbinger of a new science. A science which will place on a sound scientific basis such important subjects as psychiatry, education, sociology, criminology and penology, and even international politics. The hope is that in the future a truly scientific psychology will enable us to control the vagaries of the human mind to the same extent that the physical sciences have given us such power over our material environment. I remember as an undergraduate Dr Tennant stating that the mental sciences still awaited their Sir Isaac Newton.

The object of my paper is to show that on purely theoretical grounds this hope is vain. There is indeed a science of experimental psychology, this science will continue to grow. But as to the great expectations that the very word psychology arouses in some minds, these hopes will always remain unfulfilled. The psychological and social sciences will not transform either by power or understanding the great and terrible problems of our present discontents. For here we have to do not merely with ignorance but with the power of evil.

Before I proceed to the main arguments in support of my thesis, I would like to call to my support two eminent thinkers. I do not of course claim their authority for what I will try to prove later, but the two following quotations will perhaps not only make my thesis clearer, but also recommend it as not entirely heretical and idiosyncratic.

The first quotation is from an eighteenth-century scientist and philosopher, G. C. Lichtenberg. In one of his aphorisms he writes:

> We must not believe when we make a few discoveries in this field or that, that this process will just go on for ever. The high jumper jumps better than the farm boy, and one high jumper better than another but the height that no human can jump over is very small. Just as people find water wherever they dig, man finds the incomprehensible sooner or later.

The second quotation comes from one who can justly be regarded as the most influential thinker of my generation, Wittgenstein. Towards the end of his *Philosophical Investigations*, Wittgenstein makes this remark:

> The confusion and barrenness of psychology is not to be explained by calling it a young science; its state is not comparable with that of physics, for instance, in its beginning. (Rather with certain branches of mathematics, Set Theory.) For in psychology there are experimental methods and conceptual confusion. (As in the other case conceptual confusion and methods of proof.)
>
> The existence of experimental methods makes us think we have means of solving problems which trouble us; though problems and methods pass one another by.

I would like, indeed, to define my thesis as a logical exposition of the fact that in psychology the real problems that confront us, and the experimental methods which are being increasingly elaborated, pass each other by. And that although experimental psychology can show us new facts and confirm new hypotheses, yet in this discipline we very soon come up against the incomprehensible.

It is true enough that Hebb can 'jump' higher than James Mill. But it doesn't follow from this that a psychologist of the future will be able to jump to infinity! That he will be able, to use Hebb's own words, to 'reduce the vagaries of human thought to a mechanical process of cause and effect'.

But now it is high time that we engaged in battle in real earnest. And first a preliminary skirmish before the main thrust at the centre. I am puzzled to find one psychologist after another repeating that psychology is still a young science. For myself I find the psychological concepts which are discussed and defined in Plato's dialogues, and even more so the myths he devises to bring home to us his fundamental themes, to be a constant source of instruction. I would claim Aristotle's *De Anima* as the first treatise to deal specifically with psychology as a separate subject. (Only a short time ago I heard a Professor of Psychology in one of our senior Universities describe Plato and Aristotle as 'superstitious blighters'; and he a believer in the Rorschach test!) Aristotle's pupil Theophrastus wrote a treatise on 'Characters', and his delineation is such that we can still recognise the types he describes among our own contemporaries. But to come to something on a more serious level, I would describe St Augustine's *Confessions* as perhaps the most profound psychological analysis ever carried out.

If it be true, as indeed it is, that 'Fecisti nos ad Te,

et inquietum est cor nostrum donec requiescat in Te', then any psychology which ignores the persistent inquietude of the human soul is a shallow and superficial affair.

Then coming to more recent times. We have Descartes' *Treatise on the Passions of the Soul.* The psychological discriminations in Spinoza's *Ethics*; Locke's *Essay Concerning Human Understanding*; Berkeley's *New Theory of Vision*; Hume's *Treatise on Human Nature*; Reid's *Powers of the Human Mind*; I could continue the list but there is surely no need to labour the point. What in heaven's name possessed Hebb to take as his origin such an unimportant figure in the history of psychology as James Mill!

The psychologists I quoted at the commencement of this paper would, I imagine, reply that they were talking of experimental psychology, and in particular the introduction of measurement and mathematical statistics into the study. They would once again refer me to the undoubted fact that physics only got under way when Galileo made it a matter of accurate measurement and the construction of mathematical formulae to explain the phenomena. Similarly chemistry as a precise science owed its beginning to the demonstration of Robert Boyle that the volume and pressure of a gas could be measured and related by a mathematical law.

Now (says the modern exponent of experimental psychology) that we have begun to introduce *measurement* and *mathematics* into the science of psychology, we can indeed speak of a new science with a triumphant future. It began with Binet introducing the conception of an Intelligence Quotient which could be expressed numerically. Then Pavlov was able to measure the strength of a conditioned reflex in terms of the quantity of saliva secreted. Similarly the maze learning of rats could be quantified in terms of errors made, time taken to learn, time taken to run. And motivation could be expressed in terms of the extent of deprivation of food or water. Certainly psychology has in the last seventy years become increasingly mathematical, often requiring an advanced knowledge of statistics to be even understood.

It is a big assumption though to assume without discussion that precisely the same method which has proved so powerful in the physical sciences will be applicable to every other investigation. Aristotle, at the commencement of the treatise I referred to a moment ago, makes precisely this point. He warns us: 'If there is no single common method by which we may discover what a thing is, the treatment of the subject becomes still more difficult; for we will have to find the appropriate method for each subject.' I believe we should take this warning of Aristotle's seriously (and not dismiss him as a superstitious blighter). We should ask ourselves first, can psychology make use profitably of the same methods that have been so advantageous in the physical sciences? I emphasise the word 'profitably', for I am not calling in question the accuracy of the measurements made, but only questioning their present importance and future promise.

Let me put it this way. I have here in my hand a piece of chalk. What interests the physicist and the chemist are the properties that this piece of chalk has in common with every other piece of chalk. Its density,

DDW

its specific heat, its molecular composition, etc. No doubt this piece of chalk is in some sense unique; no other piece of chalk in the world has exactly the same markings, the same shape, the same size, but these peculiarities are of no interest to science.

But what about an individual human being? He no doubt has many properties which he shares with every other human being, and some he shares with a particular group of human beings. But to me, at any rate, what is of supreme interest is just the uniqueness of this very person, the way in which he differs from any that ever came before him or will come after. His individuality, his unpredictability, his uniqueness. In a popular work entitled *Sense and Nonsense in Psychology*, Eysenck gives a list of people whom he would classify as typical introverts, and a similar list for extroverts. His list of extroverts includes Boswell, Pepys and Cicero. I don't imagine Eysenck intended this list as more than a rough indication of what he was wishing to describe, but it will do very well for the point I am trying to bring home. It may be true that Boswell, Pepys and Cicero all have some common abstract trait in common, but that is not what is psychologically interesting or important. What interests me is that Boswell, in spite of his ludicrous vanity, his gross licentiousness, his petty mindedness, was able to write the greatest biography in literature. As Macaulay well puts it: 'Many great men have written biographies, Boswell was one of the smallest of men and has beaten them all.' Then again take the picture that Boswell has given us of Dr Johnson; it tells us little to hear that Johnson was a high church Tory, a great Latin scholar, and a learned lexicographer. What does tell us a lot are the details Boswell gives us of his conversation and repartee. Even more psychologically interesting is the deep and lasting friendship that sprang up and endured between these two very diverse characters. I am reminded here of a wonderful remark that Montaigne made when asked for a definition of friendship, 'Parce que c'est lui, parce que c'est moi.' Now do I make myself clear? In psychology what interests us, what is of deep significance, are particulars not universals. The physical sciences are interested in the universals and the mathematical relations between them. The two subjects are not comparable.

I said a moment ago that I did not question the measurements made by experimental psychologists, I did question their significance. Take for a moment the ascertainment of the Intelligence Quotient by means of one of the properly standardised tests. I do not deny that these measurements do measure some abstract ability of the individual tested, call it intelligence if you wish. But remember a remark of Janet's: he said that the most important book every written on psychology was a dictionary. Why a dictionary? Well, because a dictionary reminds us of the enormous vocabulary that mankind has found necessary to express all the different facets of personality. Just consider the word 'intelligence' and consider all the cognate words that cluster round it. Wisdom, cleverness, depth, originality, genius, clarity, docility, perseverance, and I could add more. And remember that although these are all separate words in the dictionary they are intimately blended together in the person. If I may quote Lichtenberg again:

The qualities we observe in our souls are connected in such a way that it is not easy to establish a boundary between any two of them, but the words by which we express them are not so constituted; and two successive related qualities are expressed by signs which do not reflect this relationship.

Let me then give an example from my own experience of taking the Intelligence Quotient too seriously. When I was a regimental medical officer during the war we found that one of our new recruits could neither read nor write. He was sent to the area psychiatrist for an opinion and was returned with the report that his mental age was that of a boy of twelve and a half; he was recommended for discharge. He didn't want to be discharged and it turned out that he was an expert in handling dogs and ferrets, and as we were at that time plagued with rats he was appointed official 'rodent operator' to the unit. But I needn't labour the point for I think that we are all becoming aware of the limitations of intelligence testing. I would, however, use this story for a brief digression concerning 'animal psychology'. I have for many years been an avid reader of books describing animal behaviour both in laboratory experiments in learning, and more especially the field observations made by 'ethologists'. Such a book as Thorpe's *Instinct and Learning in Animals* is one I turn to with interest from time to time. But in all these writings I find missing that which for me is the most important fact about animal behaviour; it is that all living creatures from the simplest to the nearest human are just about the most un-understandable things in the world. It is this 'un-understandability' that makes the patient watching of them such a fascination. The increasing mechanisation and urbanisation of modern times is depriving us of any close contact with the wild things of nature. This is a great psychological impoverishment. There is an aphorism of Wittgenstein's: 'That if a lion could speak we would not be able to understand him.' This one sentence says more to me than all the books that pretend to *explain* animal behaviour.

But it is time to come back to my main thesis. I want to say that the word 'psychology' is a Janus-faced word, a word that faces in two opposite directions. And that it is the fact that these are two *opposite* directions that is of the greatest importance. The first direction, and which I would claim is the original meaning of the word, occurs in phrases such as this. We might say of a great novelist such as Tolstoy, or our own George Eliot, that they show profound psychological insight into the characters they depict. Or again we would say of a historian such as Burckhardt that he had great psychological acumen in penetrating the motives behind the facts of history. In general, then, it is the great novelists, dramatists, biographers, historians, that are the real psychologists. For the sake of future clarity I am going to refer to this meaning of the word psychology as 'psychology A'.

Now the other meaning of the word psychology I shall call 'psychology B'. By psychology B, I refer to those subjects that are studied in a university faculty of psychology and are necessary to obtain a degree in that subject. The copious literature on perception in all its modalities. The numerous experiments and very diverse theories that are subsumed under the name of 'learning theory'. The various and conflicting schools of 'abnormal psychology'. Personality testing, vocational guidance, statistical method, and so on. Quite an undertaking.

But here I seem to hear the voice of my former teacher, Wittgenstein, thundering at me. 'Give examples, give examples, don't just talk in abstract terms, that is what all these present-day philosophers are doing.' So now I want to give an example of what I mean by 'psychology *A*'. It is a letter written to one of her pupils by Simone Weil. Notice that it contains nothing that one could call learning or cleverness, no attempt to be scientific or dictatorial. Yet it does contain profound psychological insight and not only for the particular individual and her immediate state, but also and perhaps even more so, for us living at the present time thirty-five years after it was written. (I only include that part of the letter which serves to illustrate my conception of psychology *A*.)

I have talked enough about myself, let's talk about you. Your letter alarms me. If you persist in your intention of experiencing all possible sensations – although as a transitory state of mind that is quite normal at your age – you will never attain to much. I was much happier when you said that you wished to be in contact with all that was real in life. You may think that they both amount to the same thing, on the contrary they are diametrically opposed. There are people who live only for sensations and by means of sensations; André Gide for example. Such people are in reality deceived by life, and as they come to feel this in a confused manner, they have only one refuge, to conceal the truth from themselves by miserable lies. The life which is truly real is not one that consists in experiencing sensations, but in activity, I mean activity both in thought and in deed. Those who live for sensations are parasites in the material and moral sense of the word compared with those who labour and create; these are the true human beings. I would add too that those who do not run after sensations are rewarded in the end by much that is more alive, deeper, truer, less artificial, than anything the sensation seekers experience. To sum up, to seek after sensations implies a selfishness that revolts me, that is my considered opinion. It obviously does not prevent love, but it does imply that those whom one loves are no more than objects of one's own pleasure or pain, it overlooks completely that they exist as people in their own right. Such a person passes his life among shadows. He is a dreamer, not one who is fully alive.

About love itself I have no wisdom to give you, but I have at least a warning to make. Love is such a serious affair, it often means involving for ever your own life and that of another. Indeed it must always involve this, unless one of the two lovers treats the other as a plaything; in that case, one that is only too common, love has changed into something odious. You see, the essential thing about love is that it consists in a vital need that one human being feels for another, a need which may be reciprocated or not, enduring or not, as the case may be. Because of this the problem is to reconcile this need with the equally imperious need for freedom; this is a problem that men have wrestled with since time immemorial. Thus it is that the idea of seeking after love in order to find out what it is like, just to bring a little excitement into a life which was becoming tedious, etc., this seems to me dangerous, and more than that, puerile. I can tell you that when I was your age, and again when I was older, I too felt the temptation to find out what love was like, I turned it aside by telling myself that it was of greater importance for me not to risk involving myself in a way whose eventual outcome I could not possibly foresee, and before too I had attained to any mature idea of what I wanted my life to be and what I hoped for from it. I am not saying all this as a piece of instruction; each one of us has to develop in our own way. But you may find something here to ponder over. I will add that love seems to me to carry with it an even more serious risk than just a blind pledging of one's own being; it is the risk of becoming the destiny of another person's life, for that is what happens if the other comes to love you deeply. My conclusion (and I give you this solely as a piece of information) is not that one should shun love, but that one should not go out of one's way to try and find it, and especially so when you are very young. I believe at that age it is much better not to meet with it. . . .

I think you are the sort of person who will have to suffer all through your life. Indeed I am sure of it. You have so much enthusiasm, you are so impetuous, that you will never be able to fit into the social life of our times. But you are not alone in that respect. As to suffering, that is not too serious a matter so long as you also experience the intense joy of being alive. What is important is that you don't let your life be a waste of time. That means you must exercise self-discipline.

I am so sorry that you are not allowed to take part in sports: that is exactly what you need. Try once more to persuade your parents to let you do this. I hope at least that happy days hiking in the mountains is not forbidden. Give those mountains of yours my greetings.

I regard this letter with its deep personal and individual message, yet also one that has a much wider implication, as a perfect example of what I have called psychology *A*. You will see that it has nothing whatever to do with the sort of studies that a degree course in academic and experimental psychology provides. So I distinguish this latter by calling it psychology *B*. Now you may say to me that there is nothing new in this distinction between a psychology which has insight into individual characters, and a psychology which is concerned with the scientific study of universal types. It is a distinction that many competent writers on the subject refer to in the Introduction to their books. I know no better account of just this distinction than what Eysenck has to say in the Introduction to his book on *Abnormal Psychology*. But my object is not just to refer to the two different meanings of the word psychology, I want to draw attention to what I have called their two different *directions*. I have the impression that most psychologists think that in time what I have called psychology *B* will enable them to be much more efficient and scientific in dealing with problems in psychology *A*. That at least is what I make of all this talk of a young science and its unlimited promise for the future. When Hebb talks of 'understanding behaviour and reducing the vagaries of human thought to a mechanical process of cause and effect', or when Eysenck states that the same laws of learning apply 'to neurotics, college students and rats', I take them to imply that the day will come when clinical insight and intuitive personal understanding will no longer be necessary, for the science of psychology will replace this primitive way of coping with problems.

It certainly has happened in the history of science that a particular and difficult skill has been replaced by a new and more accurate scientific technique. The old physicians used to pride themselves on what they called the 'tactus eruditus', the ability to gauge the patient's temperature by laying their hand on the forehead; nowadays any probationer nurse can make a more accurate reading by the use of the clinical thermometer. Similarly it used to be customary to judge the degree of anaemia by the pallor of the palpebral conjunctiva; now it is possible to estimate the percentage of iron in the blood with accuracy.

So why shouldn't a further progress in the science of psychological testing improve by means of psychology B the rough intuitive guesses of psychology A? Well, I would say the analogy is all wrong. For it is always measurement that is improved by new technique, and the importance of psychology A is that it deals with the *immeasurable*. It would only be a superficial and puerile estimation of personality that (say) took as its measure the Intelligence Quotient, the amount of dollars earned, the rank held in society, the position held in Eysenck's normal-neurotic and introversion-extroversion dimension. These are capable of being expressed numerically. But remember that most important book, the dictionary, and think of all the numerous words that we need to describe all the facets of personality. What are they? Well, here is a venerable list that will do as well as any: Love, joy, peace, longsufferingness, gentleness, goodness, faith, meekness, temperance – and if any psychologist thinks these are measurable he only shows that he does not understand the qualities the words refer to. It is hidden inwardness that is the rock over which a scientific and objective psychology will always come to grief. The truth is that we human beings are not meant to study each other as objects of scientific scrutiny, but to see each one as an individual subject that evolves according to its own laws.

So I wish to state emphatically that psychology *B*, whatever progress it makes (and I intend to discuss the proper direction of that progress later), will never replace psychology *A*, as a more accurate, a more scientific, a more efficient discipline.

I would not have considered this paper necessary if the belief that psychology *B* would one day take over the work of psychology *A* had merely been a pious hope. What does seem to me to be serious is that in some places I find the belief that the dawn has already broken. I will begin by some rather trivial examples and then go on to what I regard as more dangerous errors.

If you go into a bookshop today you will find books with some such titles as: *How to Win Friends and Influence People*, or *The Power of Positive Thinking*, or one that amused me the other day, *How to Help your Husband to be a Success*. Well, we all know that such books never helped anyone who was in serious emotional difficulties, still less anyone who was mentally ill. These books are full of harmless platitudes, but the fact that they attract a public to buy them is far from harmless; it shows a thoughtless attitude to the deeper problems of human life. It reflects a widespread error which extends even to educated people, that for every problem there is some particular *science* and some particular *expert* who can provide the necessary answer in a book.

Not so long ago an intelligent medical student came to see me. He had decided to change from the study of medicine to the faculty of psychology, for he felt that this latter study would not only enable him to cope more efficiently with his own personal problems, but also make him able to advise on the problems that other people had to contend with. I had some difficulty in persuading him that this would be a serious error to make, that he would find academic psychology a barren subject as compared with a knowledge of general medicine.

Then again a good many years ago now I was asked to give a course of lectures on 'normal psychology' to a class of students in the School of Physiotherapy. I imagine that the idea behind this request was that physiotherapists would in the course of their work have to cope with patients who needed tact and understanding in their handling. Now of course it is important that physiotherapists should have such tact and understanding, but this they will only acquire by experience and by working under one who already has achieved such wisdom. They could learn nothing from a course of lectures on 'normal psychology'. The lectures were not given.

In the medical curriculum today it is obligatory for students to attend and pass an examination in normal psychology. These lectures are usually given just after the student has completed his course in anatomy and physiology. The idea I imagine is that just as a grounding in anatomy and physiology is a necessary prolegomenon to the study of pathology and *materia medica,* so a course on normal psychology will be the groundwork by means of which the future instruction on psychiatry will be built. But the analogy is all wrong. Normal and abnormal psychology do not stand in this relation to each other. After a time I gave up giving these lectures for I felt they were an unnecessary burden on an already overcrowded curriculum.

We all from time to time in the course of our clinical work come up against personality problems where we feel out of our depth. (Perhaps we would be better psychiatrists if we felt this more often.) I notice a temptation both in my own case and that of some of my colleagues to think that such problems could be solved for me by recourse to a 'clinical psychologist'. That he by means of some superior science, some highly sophisticated tests, will come up with the right answer. The analogy that one has in mind is the way the clinical pathologist can help with a specialised blood examination, the radiologist with his experience in interpreting the skiagram, or now the extremely specialised interpretation of the electro-encephalogram. But if one looks more closely the analogy breaks down completely. For these last mentioned specialities all employ a standard technique, the facts of which I am acquainted with but lack the necessary daily practice of. But if I send a patient to a clinical psychologist I do not know what he may be subjected to. Will he be asked to interpret certain standard formless ink-blots? Will he be subjected to a series of behavioural tests? I would myself have to have enough knowledge of personality testing to know whom to trust. And maybe when I have read some of the literature on these tests and been presented with some of the clinical psychologists' reports, I will wish no longer to have recourse to this very dubious help. The clinical psychologist may assess a personality in terms of the Intelligence Quotient, the degree of introversion, the adjustment to 'reality' (though here I would want to ask what 'reality' means). I have even seen the annual earning in dollars used as a measure of 'success', but all those aspects of personality which are of deep importance escape out of the net of knowledge.

I must add a warning here though. It may well be that a particular clinical psychologist is gifted with that form of insight that I have called psychology *A*, and be able to help just because of this gift. All I am protesting against is the supposition that there exists already, or will soon be perfected, a scientific technique which will render the clinical insight gained by long experience an unnecessary acquirement. Here, now, and always, the old rule holds: 'Cor ad cor loquitur.'

There is, however, one point at the present time where 'psychology *A*' and 'psychology *B*' have already come into conflict. And this I must now discuss. There has grown up in 'psychology *B*' an immense literature concerned with what is called 'learning theory'. And some psychologists are claiming that the application of 'the laws of learning' can provide a scientific treatment for all forms of 'neurosis' and possibly some forms of psychotic behaviour as well. Thus remember Eysenck's statement:

> The laws of learning theory, to take but one example, apply no less to neurotics than to rats and college students.

Hilgard has written a comprehensive book of which the title is: *Theories of Learning* – notice the plural. For in this book he describes no less than ten different theories of learning. Osgood, in his *Theory and Method in Experimental Psychology*, has shown that many of the differences between these diverse theories are largely semantic, but in spite of this there is all the difference in the world between classical Pavlovian conditioning, Skinner's operant conditioning and the field theories of the Gestalt school. So if we are to make use of 'learning theory' in therapeutics it would be necessary to say which school of thought we support and why. As a matter of fact the behaviour therapists seem to accept some form of 'stimulus-response' theory of learning entirely, without as far as I have been able to discover any reply to the cogent arguments and experimental facts produced by the Gestalt theorists. Still less have they paid any attention to what Freud called 'Instincts and their Vicissitudes', for it is a dogma of the behaviour therapist that all neurotic symptoms are learnt in their origin, and can be cured by the application of 'learning theory'.

But it would lead us too far astray to go into the complications involved in the different theories of learning or the place to be attributed to disturbed instincts in neurotic behaviour. I would avoid such a digression by coming right away to what I regard as the central error in the whole attempt of psychology *B* to establish one unified theory of learning. For the truth is that the word learning has a great many *different* meanings and there is not one special characteristic common to all forms of learning. It may be true that we human beings sometimes form conditioned reflexes, sometimes make use of the mechanism called operant conditioning: these may lie at the basis of some fundamental habits. But that we learn in many more important ways seems to me obvious. Consider a child learning its native language at the age of three, what an inexplicable wonder is there. Skinner taught his pigeons to do many strange and unexpected things, but he never was rash enough to try to teach them to talk to him. Then later a child begins to think for himself, as the phrase goes, and it is the function of a good teacher to stimulate interest in new subjects and in a general desire for clarity and truth. These become ends in themselves. All the experiments in animal learning depend on primary or secondary reinforcement of the basic animal needs. No animal desires truth for its own sake, and yet surely that is the prime object of human education. I cannot understand how such a clear thinker as Eysenck shows himself to be, could allow himself to state that 'the laws of learning apply equally to neurotics, college students, and rats'. Eysenck, I believe, when teaching his students, wants to arouse an interest in psychology for its own sake, and for truth as an end in itself.

This enormous difference between what we mean when we talk about human learning and when the experimental psychologist talks about animal learning, becomes very manifest when we consider the great difficulty that all behaviour therapy comes up against when it tries to employ the language of animal learning to the correction of human behaviour. What is reinforcement for a human being? The rat wants its food at the end of the maze; the pigeon gets its grain when it has discovered the necessary lever to peck at; and Köhler's chimpanzee reaches his banana when he discovers how to join two sticks together. But human beings in the throes of emotional conflict are not rewarded by toys such as these. Punishment is easier, but I find, and agree with the more critical behaviour therapists, that 'avoidance therapy' is ethically unpleasant and therapeutically inefficient.

Perhaps I can clinch the matter by introducing here a little fragment of psychology A. Just two lines from the poet Aeschylus:

> Zeus has opened to mortal man a way of knowledge;
> he has ordained a sovereign decree – through suffering
> comes understanding.

The knowledge that Aeschylus speaks of here, the understanding that comes from having suffered one-self, what has this to do with the theories of learning that the experimental psychologist writes about? And yet surely it is such an understanding that is needed in psychiatry.

When I commenced the study of psychology forty-five years ago it was considered then advisable that the student should combine this subject with the study of logic, ethics, and metaphysics. But today there is a tendency to break away from this tradition and for psychology 'to go it alone'. It is considered desirable for psychology to become one of the experimental sciences and be no more dependent on a general philosophical education than, say, physics or chemistry. Now certainly the physical sciences have progressed and continue to make progress without

EDW

any need of becoming involved in disputes about first principles or ultimate objectives. But it is my belief that psychology is in a very different position, and that the more ancient tradition rested on a wise understanding. For the great difference between psychology and any other science is that the psychologist is himself part, and a very important part, of the subject matter itself. I do not see how an intelligent student of psychology could help becoming involved in all the logical problems that surround the concept of 'self'. What gives unity and continuity to his own personality. The 'I' which is always subject and never object. The logical difficulties involved in the concepts of 'inner' and 'outer experience'. These puzzlements cannot be resolved by any experimental investigations, they are prior to all experiments, they are the subject matter of logic.

If then the student of psychology must become involved in problems of logic, still more so will he come face to face with the great problems of ethics. For the psychologist like any other human being must recognise that a sense of 'oughtness', of obligation, forms a fundamental part of what being a person means. The question of what is good in itself as distinct from mere means, what is the ultimate meaning of life, the realm of ends – a psychology which excluded the enormous part that these questions have played and continue to play in the life of the mind would be put a pale abstraction from the real life of the individual. I am not of course saying that it is the function of psychology to answer such questions, I am saying that a student of psychology who does not find his chosen subject leading him away from experimental procedures to thinking about ethics is to me a strangely absent-minded thinker. For whatever we may learn from observing the behaviour of animals, whatever we may learn from experiments on perception, whatever we may learn from the study of 'individual differences', the great questions still stand over us, whence? whither? how? If you go back to a previous generation of psychologists (that is to say before behaviourism had come to dominate the scene) you will find that they realised the necessity of a comprehensive psychology saying something about this important aspect of human life. Thus William James came to write his *Varieties of Religious Experience*; Freud could not rest until he had written *Totem and Taboo, The Future of an Illusion,* and *Moses and Monotheism*; Janet too, in his two big volumes *De l'Angoisse à l'Extase,* is concerned to give an account of the ethical and religious experiences which have played and will continue to play a dominant role in human behaviour.

It is curious to say the least that a generation which prides itself on the frankness with which sexual problems are handled, should seem almost embarrassed by any reference to guilt, sin, death and judgment. Yet thoughts about these concepts must play a part in the thinking of anyone who is fully alive.

It is my belief that many turn to the study of psychology because of the pressure of these great problems, which seem to be part of the phenomena with which the instruction should deal. They will feel a sense of frustration if their teachers have nothing to say on such matters. If they find psychology confining itself to averages and statistics and experiments on rats running in mazes. Then on the other side too I think it is valuable for the student of philosophy to have one of the experimental sciences as part of his curriculum; and experimental psychology is eminently suitable to play this role. Experimental psychology, besides giving good examples of the difficulty of devising crucial experiments, is replete with concepts that are in urgent need of dialectical development. All in all, then, I regret this tendency for the faculty of psychology to break away from its previous companions and become an independent study.

I am afraid I may be giving the impression that I attach very little importance to the study of experimental psychology. If so, I would now like to correct that imputation and say where I think the real importance of these experiments lies. I believe that experimental psychology has made and will continue to make very significant contributions to the study of neuro-physiology. Pavlov always described his work as 'the *physiology* of the higher nervous activity', and he eschewed the claim to be a psychologist. Yet Eysenck has called Pavlov the greatest of experimental psychologists. There is only a verbal difference here. It has interested me to see over the years how Eysenck's own books have taken on more and more a physiological terminology to replace a previous psychological one. In his first book we had the twin dimensions of extroversion–introversion, and neurotic-normal. Now these dimensions are defined in physiological terms. The extrovert is one who finds difficulty in establishing positive conditioned reflexes, and can easily inhibit those already formed; whereas the introvert quickly establishes positive conditioned reflexes but finds the subsequent inhibition of those so formed harder to achieve. The neurotic-normal dimension is defined in terms of the degree of reactivity of the autonomic nervous system. Psychologists may dispute as to whether this theory is yet established, but the direction that the investigation has taken seems to me the correct one.

I mentioned earlier that Hebb's book on the 'organisation' of behaviour has had considerable influence. This I believe is due to the fact that Hebb brings out clearly how experiments in psychology have forced an elaboration of previous neurological constructs which are too simple to account for all the facts. For instance Lashley's search for the engram produced evidence that made all simple 'stimulus-response' connectionship, such as Pavlov, Thorndike, and Lashley himself believed in, impossibly simple. Similarly the perceptual phenomena that the Gestalt school drew attention to have made us revise considerably a too simple theory of sensory representation in the cerebral cortex. Lichtenberg in one of his aphorisms says that 'Materialism is the asymptote of Psychology'. I am not sure that I understand what he meant. But I would certainly say that neurophysiology is the asymptote of experimental psychology. The more rigorous experimental psychology becomes the more it will need to translate its findings into physiological terminology. Hebb seems to me to make this very point admirably in his book. But then he goes on to say:

Modern psychology takes for granted that behaviour and neural function are perfectly correlated, that one is completely caused by the other. There is no separate soul or life force to stick a finger now and then into the brain and make neural cells do what they would not otherwise.

Now while I would agree that *some* forms of behaviour are correlated with neural function and that in this field experimental psychology and neuro-physiology can with profit co-operate, I would most emphatically deny that *all* behaviour is so correlated. Nor would I conceive it a necessary alternative to believe in a 'soul or life force sticking its finger into the brain'. Why should there not be some areas of behaviour which just have no neurological counterpart? Let me put it this way. I feel hungry, and I have good experimental evidence to think that this particular sensation is correlated with a lowered blood glucose level and peristaltic contractions of the stomach. I study the menu and select certain foods; now there is some evidence from experimental psychology that animals deprived of certain necessary dietary elements will instinctively select those foods which will correct their deficiency. It may be true that human beings select their food on such a physiological basis, though this has never been proved, and the evidence of man's dietary indiscretions are all against it. After the meal I feel a desire to listen to some music; maybe with the further development of the electro-encephalogram this desire may be found to be correlated with some particular pattern of electrical wave formation. But then from my records I select, say, Bach's fourth Brandenburg Concerto. Now it seems plain nonsense to me to say that this individual selection is *physio-*

logically determined. I may or may not be able to give *reasons* for my choice, but it does not make sense to ask at the same time for the *cause* of my choice. Hebb, for example, in his book gives the evidence and the reasons which have led him to his particular theories; I imagine he would be justifiably annoyed if we then asked him to give the causes of his beliefs. For if it is not possible for us to choose between truth and error, between right and wrong, then the whole possibility of scientific discussion is reduced to an absurdity. The very possibility of speech, of intelligent discourse, of well reasoned books, depends on the certainty that a very large and important part of mental life is not determined and is not correlated with specific neural function. This certainly does not imply that a finger is thrust into the brain to compel neural cells to do what they otherwise would not. Nothing is compelled because nothing is correlated. You couldn't carry on a discussion with a tape recorder where everything is correlated and compelled; you can carry on an argument with another human being because he is able to *choose*. If I may say so, Hebb's error here is an excellent example of the point I was trying to make about the importance of a student of psychology having an acquaintance with logic. For any modern student of logic will immediately remember Wittgenstein's aphorism that 'Belief in the causal nexus is *superstition*.'

Psychology a new science with wonderful promises of future power and mental transformations: psychology determining through the study of animal behaviour the laws of learning which apply equally to 'neurotics, college students and rats'. So then I would end with this quotation written now nearly 2,500 years ago. At the end of his dialogue, the *Philebus*, Plato puts into the mouth of Socrates these words. Socrates speaks:

> Our discussion then has led to this conclusion, that the power of pleasure takes the fifth place. But certainly not the first, even though all the cattle and horses, and every other living creature seem to imply otherwise by their pursuit of enjoyment. Those who appeal to such evidence in asserting that pleasures are the greatest good in this life, are no better than augurs who put their trust in the flight of birds. They imagine that the desires we observe in animals are better evidence than the reflections inspired by a thoughtful philosophy.

A new science?

CHAPTER 3

CONCERNING BODY AND MIND

WHAT an odd situation the soul is in when it
reads an investigation about itself, when it looks
in a book to find out what itself might be. Rather like
the predicament of a dog with a bone tied to its
tail – said G.C.L., truly but a little ignobly.

LICHTENBERG

Ladies and gentlemen,

When I had completed this paper I had doubts as to
whether I should read it to you. For our Society is
properly concerned with those particular problems of
diagnosis and treatment which are daily met with in
our work. And much that I have written here will seem
at first to you as so much barren metaphysics. I don't
believe that it is barren metaphysics; indeed I hope
to show you before I conclude that the thoughts here
developed have important practical consequences, if
not for the details of our work, at least for the general
ethical background against which our work must be
carried out.

I do not see how anyone can practice the profession
of a physician, still less of a psychiatrist, without soon
coming face to face with the deepest philosophical
problems as distinct from scientific ones. Problems

which by their very nature cannot be answered by any future development of scientific discovery, but require an altogether different method of investigation. And if we do not at some time or other stop to ponder these problems, the very facts with which we come face to face will necessitate some answer in our actions. And so I venture to read this paper. But if to some of you it seems out of place, I can only assure you that I share in your disquietude.

Some months ago I was reading a translation of one of Pavlov's Wednesday morning conferences. Every Wednesday Pavlov used to meet his students and assistants, and other visiting scientists, for a general and informal discussion about any topic connected with his work in neuro-physiology. A stenographer kept a record of these conversations and they have now been translated in part. I have found some of them very interesting reading.

On Wednesday, 19 September 1934, Pavlov arrived at the conference with his bushy whiskers fairly bristling with indignation. He had been reading a book by the famous English physiologist Sherrington. In this book Sherrington had used the words 'if nerve activity have relation to mind'. Pavlov was so shocked that any physiologist should doubt the absolute dependence of mind on brain, that he thought he must have read a mis-translation. He got a friend who had a better knowledge of English to check the words for him: but there they were in all their scandal – 'if nerve activity have relation to mind'.

How is it possible, said Pavlov, that in these days any scientist, let alone a distinguished neuro-physiologist, could for one moment doubt the com-

plete dependence of the mind on the healthy functioning of the organic nervous system? After some discussion he summed up his conclusion in these words:

> Gentlemen, can anyone of you who have read this
> book say anything in defence of the author? I believe
> this is not a matter of some kind of misunderstanding,
> thoughtlessness, or mis-judgment. I simply suppose
> that he is ill, although he is only seventy years of age,
> that there are distinct signs of senility.

For Pavlov, then, after thirty years of studying conditioned reflexes, the complete dependence of mind
and brain was axiomatic.

Mind dependent on brain. I suppose that we who
use daily physical methods of treatment in psychiatry,
and too often see the disastrous effects that organic
disease of the brain can produce both on intellect and
character, would feel inclined to agree with Pavlov.
To think of the mind and its activities as in some way
the product of that complex tissue we call the brain.
But I would also guess that if we were asked we might
well find ourselves at a loss to say how exactly we
conceived this precise dependence.

In the previous century that great biologist T. H.
Huxley, lecturing to an audience, told them: 'The
thoughts to which I am giving utterance, and your
thoughts regarding them, are the expression of molecular changes in that matter of life which is the source
of our other vital phenomena.' Is that clear to you?
I must say that I can attach no clear meaning to it.
Sherrington recalls that as a student in Germany the
Professor put one of the Betz cells from the cerebral
cortex under the microscope, and labelled it 'the

organ of thought'. A few days later a tumour of the brain was being demonstrated in the pathology department and one of the students asked: 'And are these cells also engaged in thinking, Herr Professor?' Now this I think was a really witty remark. For it made a piece of concealed nonsense obvious nonsense.

But to come to something written nearer our own time. Professor J. C. Eccles published in 1953 a book entitled *The Neuro-physiological Basis of Mind*. Professor Eccles is a recognised authority on neurophysiology; let us hear what he has to say.

In the Introduction to his book the Professor states his programme as follows:

> As indicated by the sub-title of this book, 'The Principles of Neuro-physiology', the scope may be described as covering the whole field of neurophysiology. The reactions of the single nerve or muscle fibre, the reactions of the single neurone, the reactions of the simpler synaptic levels of the nervous system, the plastic reactions of the nervous system and the phenomena of learning, the reactions of the cerebral cortex, and finally the relation of the brain to the mind. Broadly speaking it is an attempt to see how far scientific investigation of the nervous system has helped us to understand not only the working of our own brains, but also how liaison between brain and mind could occur. As such it tries to answer as far as is present possible some of the most fundamental questions that man can ask. What manner of being are we? Are we really composed of two substances, spirit and matter? What processes are involved in perception and voluntary action? How are conscious states related to events in the brain? How can we account for memory and the continuity of mental

experience which gives the self? How is that entity called the self interrelated to that thing called the body? Descartes failed to answer these questions because his science was too primitive, and his dualistic inter-actionist explanation has consequently been discredited. The remarkable advances that have been made possible largely by electronic techniques, now make it worth while to answer these questions in at least some of their aspects.

This is surely a most ambitious programme that Professor Eccles has set himself. And if it really is true that recent advances in neuro-physiology aided by electronics can tell us, to use his own words, 'what manner of being we are', then surely this subject and these techniques are the most important science that anyone could choose. We would be wise to leave aside all other studies for this. But before we make such a serious decision it might be wiser first to see what conclusions Professor Eccles himself has come to on these weighty matters. To see indeed whether he has in any degree at all been able to fulfil the task he has proposed himself.

The first 260 pages of his book are taken up entirely with the early part of his programme. The structure, biochemistry and electrical phenomena of the individual neurons and synaptic junction; and then the more general anatomical organisation and histology of the cerebral cortex. Here as one would expect there is much original work of great interest and ingenuity. It is only in the last twenty-six pages of his book that he passes from positivistic natural science to more speculative matters. He starts his final chapter as follows:

We now come to the problem posed at the beginning of this book, which may be covered by the general question: 'who are we'? The answer to this question is according to Schrödinger 'not only one of the tasks of science, but the only one that really matters'.

But surely we don't need a distinguished physicist, or even a Professor of neuro-physiology, to tell us how important this question is. Centuries ago a certain unknown Greek wrote over the door of the temple of Apollo at Delphi γνῶθι σεαυτόν, 'know thyself'. This injunction 'know thyself' did not mean know your own personality and peculiar idiosyncrasies, but know what it means to be a human being, what is the nature of men as such: the very question indeed that Professor Eccles mentioned in his Introduction – 'what manner of creature are we?' Let us then look first at what he as a neuro-physiologist has to tell us. He writes:

The usual sequence of events is that some stimulus to a receptor organ causes the discharge of impulses along afferent nerve fibres, which after various synaptic relays, eventually evoke specific spatio-temporal patterns of impulses in the cerebral cortex. The transmission from receptor organ to cerebral cortex is by a coded pattern that is quite unlike the original stimulus, and the spatio-temporal pattern evoked in the cerebral cortex would again be different. Yet because of this cerebral pattern of activity we experience a sensation (more properly the complex constructs called percepts) which are projected outside the cortex, it may be to the surface of the body or even within it, or as with visual, acoustic and olfactory receptors to the outside world. However the only necessary condition for an observer to see colours, hear sounds, or experience the existence of his own body, is that appropriate patterns of neuronal activity should appear in appropriate regions of his brain, as was first clearly seen by Descartes. It is immaterial whether these events are caused by local stimulation of the cerebral cortex or some part of the afferent nervous pathway; or whether they are, as is usual, generated by afferent impulses discharged by receptor organs. In the first instance then the observer will experience a private perceptual world which is an interpretation of specific events in his brain. This interpretation occurs according to conventions acquired and inherited, that, as it were, are built into the micro-structure of the cerebral cortex, so that all kinds of sensory inputs are co-ordinated and linked together to give some coherent synthesis.

What an amazing state of confusion for an intelligent man to have arrived at! Of course something similar has been said many times before. It seems to be a fatal pitfall for anyone who is too preoccupied with the details of sensory perception. Thus Professor Eccles is able to quote such a distinguished neurologist as Sir Russell Brain in support of his conclusion; he quotes Sir Russell as saying: 'Mental experiences are the events in the universe of which we have the most direct knowledge.'

Now I want to do three things. I want first to bring out clearly the confusions and inconsistencies in such a view as Professor Eccles has here put forward. Secondly to show how easy it is for such a confusion to arise, how easy it is for anyone of us to find ourselves thinking along these lines, and thirdly and most important, how to get this sort of confusion out of our system once and for all.

I say that Eccles's theory is both inconsistent and confused. Throughout all his account he is quite certain of both the meaning and truth of certain statements he makes. His final theory is in fact a conclusion drawn from these data. He starts off by using the phrase 'the usual sequence of events'; how does he know that this is the usual sequence of events? He mentions stimuli acting on receptor organs. But if, as he states, the observer experiences 'a private perceptual world', how is it that he can even begin to talk about external stimuli and receptor organs? Eccles is certain that the coded pattern in the afferent nerves and that evoked by them in the cerebral cortex is quite unlike the original stimulus. But then how does he know what these original stimuli were like?

To put it briefly then, if, as Eccles asserts, the only necessary condition for an observer to see colours, hear sounds, etc., is that appropriate patterns of neuronal activity should occur in appropriate regions of his brain; then I do not see why or how we should ever come to believe in, or even understand the meaning of, still less logically infer, the existence of a real objective world outside. And if, furthermore, 'seeing' is taken to be the same thing as 'experiencing a private perceptual world', why should we even believe in neuronal activity in the brain itself?

Eccles is vaguely aware of this difficulty. And in fairness to him, and because it brings out the confusion even more clearly, we should consider his attempted solution. He writes:

> We report our mental experiences to others and find
> they have like experiences to report to us. Such proce-
> dures serve to assure us that our private experiences
> are not hallucinations, or more strictly we may say
> that hallucinatory experiences are discovered by this
> procedure. We may conclude then that our mental
> experiences cannot be rejected as hallucinations, nor
> is solipsism a tenable explanation. Mental experiences
> are reported by all human beings with whom we take
> the trouble to communicate at the appropriate level.

But here once again Eccles is assuming the very truth
he wants to regard as a justifiable inference. For if we
really begin by being shut up in our own private perceptual world, how do we ever come to know that
there are other observers? How could we have ever
come to talk to them and to share a common language? If the only reason for regarding our sensory
experiences as not hallucinatory is that we hear other
voices, why should not these voices also be hallucinations (after all we know only too well that hearing
voices is the most usual form of hallucination).

We have seen a very distinguished and competent
experimental neuro-physiologist writing a lot of nonsense. But this particular brand of nonsense is very
close to all of us. We have no cause to flatter ourselves
at his expense. Let us go over the ground again and
see how easy it is to get into the same sort of confusion. For only by looking at this problem from
every angle will the essential and necessary clarity
appear.

By accident, say, I touch the top of a hot stove and
sustain a painful burn. Now we know that if, for
instance, I had a disease of the spinal cord such

FDW

as Syringomyelia, I would see both the stove and the blister on my finger but would feel neither the heat nor the pain. And so we make a distinction: the stove, my finger, the blister, these are real external things; but the feeling of heat and pain is subjective and in the mind of the percipient.

Or again I come into a room where there is a vase of roses, and I enjoy the perfume from them. But if my nose is stuffed up by a heavy cold, the smell is lost to me. So once again we say that although the bowl and the roses which are before our eyes are real enough, the smell and the pleasure I take in it are somehow in the mind of the observer. And we attribute this smell to minute particles given out by the roses which cause a chemical reaction on the nerve endings of the olfactory nerves in the nose. Here notice we have had to introduce something purely hypothetical, the minute particles.

Once more, I am listening to a concert on the radio. I have to leave the room. The radio set we are sure remains there just as it was, and also the sound waves in the air set up by the vibrations of the loud-speaker. But the music and all that it had meant to me, that surely leaves the room when I do. The symphony we would be inclined to say is the effect of the waves of air impinging on my eardrum, moving the ossicles of the middle ear, and transmitting nerve impulses to the auditory cortex via the eighth cranial nerve.

But now notice what we have been doing all along. We have been drawing a radical distinction between the sense of sight and what it reveals, and all the other sensory organs. At an unreflective and 'common sense' level it would appear as if the senses of touch, smell, hearing and taste were all dependent, as Eccles expresses it, on 'receptor organs'. But to the naïve observer the eyes are almost forgotten when in use. We are inclined to think of the eyes as in some sense a pair of windows through which we look *out* on the external world; not as organs of sensation *intervening* between us and the world of real things. But if we are to base any conclusions on physiology, the eyes are as much receptor organs as any one of the other senses. There is the transparent cornea, the lens with its possible defect of accommodation and transparency, the highly complicated nervous and photo-chemical structure of the retina, the optic tract with its decussation, the synaptic relay at the level of the geniculate bodies, and finally the radiation to the calcarine fissures of the occipital lobes. No wonder then that the physiologist finds himself in the position of saying that 'In the first place the observer will experience a private perceptual world which is an interpretation of events in his own brain.' We are back again at that complete subjectivism which we saw contained so much contradiction.

Let us go over the ground once more and see if we can track down this error to its source. We, like Professor Eccles, have been basing our conclusions on the certainty that we know a good deal about the anatomy of the sense organs and the nervous pathways to and in the brain. How did we acquire this knowledge? Well, think back for a moment to that day in the school of anatomy when you first removed the calvarium of the skull, and saw before your own eyes that marvellous structure the human brain. And then at once we had to get down to the laborious task of memorising the names of all those many, many fissures and lobes, of the areas of grey matter and white matter, and the cranial nerves which proceeded from them. But scarcely had we completed this task than the physiologist was upon us. This structure which you see with the naked eye is not, so he told us, the real nervous system; just look here beneath the microscope, see these different nerve cells, Betz cells, Purkinje cells, neuroglia, axions, dendrites, and the multitude of synaptic junctions between them. This immense neuronal network, every brain containing many more cells than there are inhabitants of this earth, this neuronal network is the real nervous system.

Not at all says the biochemist and geneticist. The dead and stained specimens on the microscope slide are largely artefacts; we know that the real nervous system is a living, growing, constantly changing thing, the seat of most complex chemical transformations and electrical phenomena which continue both by day and by night. So once again we had to get down to the task of learning the Eberden-Meyerhof process for the anaerobic utilisation of glucose, and the Krebs cycle for the oxidation of pyruvic acid. Do you remember now those benzene rings with their long side chains of carbon, hydrogen, and oxygen? And now more recently the geneticists have produced the double helix structure of the fundamental substance D.N.A. and seen in this, so they say, the key to the genetic code. Here, of course, we have passed out of the field of even microscopical vision. The various molecular structures can only be shown to us by drawings on the blackboard. And, if you were like me, I imagine that you could not help thinking that

these macroscopic patterns were a reproduction of a
similar one in the infra-microscopical world. I am sure
the geneticist really *believes* in his double helix. And
then as you placed the various atomic symbols in
their correct place in the structural formula, you
thought of this *C* representing a real discrete particle
of carbon, and this *S* as representing a real discrete
particle of sulphur, and so on. In fact you thought, to
use a phrase of the great Sir Isaac Newton, of 'little
particles of matter so hard as to be indivisible'.

But now if an atomic physicist had come your way
he would indeed have laughed at you. 'Little hard
particles of matter? My dear fellow, don't you know
that we exploded that theory long ago; one fine day
over Hiroshima.' The ultimate and real constituents
of matter are the fundamental particles, protons,
electrons, neutrons, and now the less stable ones that
keep coming on, neutrinos and anti-neutrinos, mesons,
and pi-mesons, and I don't know what else.

To me it is never clear when these are spoken of as
particles what they are particles *of*; and when these
particles are said to pass from one orbit to another
without passing through the intervening space then
my mind gives up. Merciful heavens, what has happened to that nice solid brain we saw in the anatomy
room? Three times now we have found ourselves
ending up in a morass of confusion. There must be
something radically at fault in our thinking, some
original sin that has led us into repeated error. I
believe this is so, and I want now to try to show this
error to you. But I am in some doubt if I can do so,
for the difficulty is not one of explaining something
very complicated and profound, but of showing the

immense importance of something so simple that it
continually escapes our notice. However, let me at
least try.

You remember Professor Eccles's original programme. He said he was going to investigate the whole
field of the nervous system. Not only the structure
and organisation of particular neurones and their
synapses, but also 'the working of our brains', 'how
liaison between brain and mind could occur'. He was
going to investigate 'the process of perception', 'to
give an account of memory and of that continuity of
experience that gives the self', and finally 'decide what
manner of creature we really are'.

'Investigate'. This is indeed a great word. We live
in an age of investigation, when everything is investigated, both in the heavens above and the earth
beneath and the waters under the earth; often with
great interest and to the relief of man's estate. So
what could seem more natural than to investigate perception and memory and the true nature of the self 'in
the same way'. There is no more dangerous phrase in
philosophy than that one 'in the same way'. Are you
sure that this same way is still open? Perhaps there is
a limit to what can be investigated by science. One
of the main tasks of philosophy is to show the limits
of what at first sight seems limitless. So let us enquire
more closely into the real nature of scientific investigation. That means taking a concrete example and
seeing what we really *do* when investigating. The
patient, say, has a temperature of 102°F., and I say
to Sister we must investigate this. And so we go
through the routine procedure of inspection, palpa-
tion, percussion, and auscultation, and then proceed

to certain laboratory and X-ray examinations. But
now notice what we are completely dependent on,
what we assume as perfectly valid, in such investigations. We depend entirely in every one of these procedures, even the most recondite laboratory ones, on
perception and memory. Sight, touch, hearing,
memory, language, these are the *instruments* of
scientific investigation. Therefore they themselves
cannot in turn be investigated. If I may use a crude
metaphor; I can look at any object on earth or in the
sky through my telescope, except the telescope itself.

It therefore does not make sense, it is really a piece
of concealed nonsense, when Eccles proposes to investigate perception, memory, the self, and in general the
relationship between mind and body. We can indeed
investigate in more and more detail the *anatomy* of the
sense organs; but the 'nature of perception', 'the
liaison between mind and brain', 'the transition from
nerve impulse to consciousness' – investigation makes
no sense here. I am not saying that these matters are
so complicated that we cannot attain unto them. That
would only be a challenge to try harder. Nor am I
saying that these things are so commonplace that we
can take them for granted. But I am saying that
however much we learn concerning the physiology of
the eye and optic tract this will never explain how
seeing is possible. Perhaps someone would like to
interrupt me here by drawing my attention to the vast
literature that already is in being concerning the
psychology of perception, memory, language. I think
here particularly of such books as Vernon's on *Visual
Perception*, Broadbent's book entitled *Perception and
Communication*, and Bartlett's work on *Remembering*.

How then can I state so emphatically that perception, memory, language cannot be investigated? But now notice in these books that the psychologist is largely concerned with experiments on subjects other than himself. He has to take as given the seeing, hearing, remembering, describing, which are his own observations; these form no part of the investigation. It is possible for a psychologist sometimes to act as his own subject; one thinks at once of Ebbinghaus and his laborious work on learning and forgetting strings of nonsense syllables. But here again the same dichotomy occurs. When Ebbinghaus came to write his book, the memory of these self-experiments, the interpretation of his own notes made at the time, these were fundamental data and were not themselves in need of investigation. All I am saying is that in every investigation there will always be that which is not itself investigated; in every experiment there will be data which are not the result of experiment; in every enquiry there will always be that which is not enquired into.

This for me is such a simple yet important and far-reaching truth that I would come to the same point by another route. Suppose you showed a clock to one who had never seen such a mechanism before. You could explain in detail to him the working of the mechanism, and the use of the machine. But now imagine one born stone deaf; he could carry out a complete dissection of the outer, middle, and inner ear, and could become a master in the cytology of the auditory cortex; and nothing in all this would ever explain to him what hearing was like. Similarly one born blind could by reading Braille answer correctly all questions concerning the structure of the eye and the optic tract, but would never come to an understanding of what was meant by sight. Then again I could give an account of how I came to learn a foreign language; but who among us could describe how he came to speak his *native tongue*?

I said that these were such simple truths that their importance is too easily overlooked. It was forgetting these truths that led us into all the confusions of complete subjectivism. Simple truths, yes, but not platitudes. If we were to lose our sight we would indeed bemoan our fate; should we not then sometimes pause and wonder at the miracle of sight? I want to say that every time you open your eyes a miracle occurs. If we should become deaf, think what we would lose in the way of friendly communication and intelligent discussion. So should we not wonder at, and be grateful for, the miracle of hearing? Every time you wake up in the morning and return to consciousness a miracle occurs.

Much speculation is going on as to the nature of the memory trace and how it is to be explained at the physiological level. A perpetuating cycle of activity in a complex of neurones? A facilitation at synaptic junctions? A molecular coding by the complex molecules of R.N.A. or D.N.A.? But whatever comes of these speculations it will still be true that:

> When to the sessions of sweet silent thought
> I summon up remembrance of things past

a miracle occurs. It is correct to speak of progress in science and in our scientific understanding of the world. May that progress continue. I hope that nothing that I say in this paper will be taken as an 'attack on science'. The greater part of my life has been taken up with an attempt to think scientifically about the problems of mental pathology, their cause and their treatment. But I am sure it is an error which is disastrous to our philosophy if we forget this great truth: that however much the realm of what is explained is extended, the realm of the inexplicable is never reduced by one iota. It would seem from much that is written nowadays that perhaps in the far distant future everything will be explained and controlled by scientific understanding. I have been trying to emphasise that this, thank God, will never be the case. Those fundamental data which we use in giving explanations: perception, memory, language, these remain for ever in the realm of the inexplicable. In an age such as this in which technological sophistication increases daily, there is a great danger that we lose the precious gifts of wonder and gratitude for the common and simpler foundations of our being.

Eccles, you will remember, spoke of the rudimentary nature of Descartes' physiology, making it impossible for him to solve the problem of the relation of mind to body. Perhaps you may have thought that in concentrating my criticism on a book published in 1956, I was already a bit out of date; after all there have been remarkable discoveries in neuro-physiology in this last decade. But my choice was deliberate. Professor Eccles was in no better position than Descartes, and we or any subsequent generation will not be in any better position than Professor Eccles, to solve a problem to which the notion of a 'solution' does not make sense. I have no need to fear that tomorrow some new discovery will invalidate everything that I have written here. 'In the idea now is always.'

At the risk of being prolix I am going to give two further examples from other writers of the type of misunderstanding I am concerned to eliminate from our thinking. (1) Suppose you take up such an excellent book as Ranson's *Anatomy of the Nervous System*. And suppose you want to learn the sensory pathway from the tips of the fingers to the cerebral cortex. You learn the details of the end organs in the fingers, the sensory nerves in the arm, their rearrangement in the brachial plexus, their separation from the motor nerves and entry into the cord through the posterior nerve roots. Then the arrangement of the sensory tracts in the spinal cord, their decussation at the level of the brain-stem, their relay to the various nuclei in the thalamus, and the connections to and from the cortex to the thalamus. All this can be described in positivistic language and we can if needs be verify each statement in the anatomy room. But then you find, even in Ranson, a reference to the question at what point the nerve impulse 'enters' consciousness. Notice the complete change of language here; this is no longer descriptive and verifiable, but metaphorical and speculative. In everyday language if we use the word 'enter' we imply a threshold both sides of which can be observed and entering means passing from one side of this threshold to the other. But in this sense, the common everyday sense, of 'entering', you cannot speak of anything entering consciousness. For consciousness has no boundary, no threshold which can be observed. If it had then there would have to be a third form of consciousness which was conscious of both what was conscious and what was not yet so. This is obvious nonsense. Consciousness is not just one of the many things we are conscious of: the mind has no particular place in nature.

(2) In a book recently published I found the following sentence:

> Some thoughts we keep to ourselves. But man being a gregarious animal seeking companionship and co-operation with his fellows, naturally wanted to pass on a great many of his thoughts. This led him through countless ages of endeavour to develop a means of communicating thought. Laboriously he built up language.

But this is surely nonsense. First thoughts and then the gradual development of words in which to express them! But can any of us think without already using words to ourselves? Thinking and language are not separable in this way. And what does 'laboriously build up' mean here? Nowhere in the world, nor in the study of extinct languages, do we find gradations of language, but only different languages. Indeed both primitive and extinct languages are often of the greatest complexity, and 'progress' is often in the nature of simplification and reduction in the vocabulary. To talk of countless ages of endeavour is to invoke a *deus ex machina* here that just won't work. I want to say that the existence of language, and the development of the ability to speak in a child is a miracle, something that the notion of explanation as to how it came, and comes to be, does not make sense. It is something indeed for us to wonder at and be thankful for.

If Pavlov had been listening to what I have said so

far I am sure his patience would have passed all bounds, and he would have been convinced of my advanced senility. And perhaps your patience is becoming exhausted too. Surely you would say that if our sight is troubling us we go to an ophthalmologist and he by his knowledge of the anatomy and physiology of the eye and optic system may be able to correct our vision. Similarly if deafness troubles us we go to an otologist. We would not think much of such specialists if they told us that seeing and hearing were miracles. Surely we must then say that sight and hearing are dependent on the physical structure of the sensory organs and their nervous connections with the brain. Once more, in the last two decades the power of physical methods of treatment in psychiatry have been increasingly demonstrated. How then can I dare to criticise one who states that mind is dependent on brain?

Now I need hardly say that I accept as much as anyone the physical treatments of the ophthalmologist, the otologist, the psychiatrist, etc. All that I am criticising is the vagueness and the many misleading interpretations which that phrase 'dependent on' can give rise to. I want to fix a more precise meaning to that phrase, to determine its *limits*. I have tried to show how much of confusion and error comes into our thinking if we do not fix and determine these limits. But more than that only, I want to make 'wonder secure'. There is a danger, with the ever increasing development of natural science, its powerful applications, and its inevitable specialisation, that we come to forget the realm of the inexplicable. 'The mysteriousness of our present being.'

At the common-sense level we found ourselves tempted to speak of an outer world revealed by sight and an inner world of feeling, hearing, remembering, etc. I hope I have been able to make you see the misunderstandings that are contained in this unreflective use of the words 'outer' and 'inner'. The words compelled us to picture a boundary which had to be crossed and yet somehow we were to be conscious of both sides of this boundary. There is no such boundary. Experience is experienced as one continuous whole. What we see, the distant hills and their colouring, the sound of a bird calling, the smell of the pine trees, the feel of the sand beneath our feet, and the memories of previous visits to this place, and the pleasure which accompanies these things, this is all given as one and undivided. For various practical necessities we break up this undivided whole, and attend now to this aspect and then to that. If I am doing anatomy it is sight that must be my guide. But if I am listening to music it may be well to close my eyes. These are matters of expediency and depend on a deliberate shift of attention. Now the advances in natural science have been due to a wise and deliberate selection of certain aspects of the total given whole, and the ignoring of others. The division of qualities into primary and secondary was a great discovery in methodology, not a metaphysical discovery. For instance I use spectacles to overcome my astigmatism, but if I am interested in the physiology of the eye it is these very distortions that interest me. Both the corrected vision and the distorted vision are equal in their ontological status, they both belong to that which is real. Everything that comes to us by way of the senses is part of reality and worthy of attention at times, what aspect we choose to study in detail is a matter of choice.

There is a flash of lightning and a clap of thunder. We are all startled by them and make comment. Here surely you would say is an 'outer experience'. The thunder has made me feel nervous but I manage to conceal this feeling from the others, here surely you would say is an 'inner experience'. Of course I am not denying such a familiar distinction as this. I am denying nothing. I am pointing out the very real dangers and confusions that those words 'outer' and 'inner' produce in philosophy. They force on us a picture of reality, of the mind and its place in nature, which is of no use at all.

The distinction between seeing the flash of lightning and my feeling of fear is not that the vision lies on one side of a barrier and my feeling on the other. For me there is no barrier between them, they are together. We have learnt by experience that if there is lightning others see it too (but remember not always); we have also learnt by a long process of training to conceal our emotions. A small child cannot conceal its terror, such an emotion is as much 'outer' as anything seen; it is *seen*. I sometimes have a high pitched ringing noise in my ears, this I constantly mistake for the telephone ringing; only experience has taught me that I really have *Tinnitus aurium*. I know now that the sound is 'inner', but at one time it was 'outer', and it has not changed its quality.

Once we have grasped the necessary limitations that must be imposed on the words 'inner' and 'outer' when used in epistemology, then the confusion we got into when describing the different conceptions of the brain can be disentangled. The anatomist describes the brain which he sees with his naked eyes. The physiologist describes the stained specimens he sees under the microscope. The biochemist describes those experiments which have led him to postulate such and such a molecular structure. The pure physicist has his own complicated apparatus for investigating the structure and constituents of the atom. But once again I must insist every one of these investigators depends in the last resort on sensory perception, memory, language; these are the tools with which he investigates and whose validity he has to assume. He cannot in turn investigate these.

No one of the pictures that these various investigators build up to direct them in their work has any claim to priority over the other. All are necessary for a full knowledge of the subject. A radiologist trying to locate a brain tumour from the appearance of his X-ray will necessarily use the gross terms of the anatomist. The neurologist trying to account for an area of anaesthesia or paresis, will be guided by the knowledge he has of neuronal structure, and will speak in terms of nerve centres and nerve tracts. The expert in mental deficiency is being increasingly helped by the development of the biochemistry and genetics of the brain. And finally it is the pure scientist who has over the past centuries provided us with the microscope, the chemical stains, the X-ray apparatus on which these other investigators depend. At no level of investigation can we say, 'Ah! now we have reached the real thing in itself; before, all that we were concerned with was mere appearance.'

*   *   *   *   *

And yet, and yet; when Professor Eccles promised to answer that question 'what manner of being are we?', did not this arouse our interest in a deep and serious sense?

We have seen, I hope, in the previous section, that this is a problem that no empirical investigation can ever answer. Schrödinger was right in describing this as the most important of questions, but then he went all astray when he added 'the most important question that *science* can ask'. It is not a problem in natural science; neither neuro-physiology, psychology, or any other empirical investigation can help us here. And yet we must both ask and give an answer at once. The fact that we have to live and make decisions demands an answer.

It was a Greek who first posed this question in words. Suppose then that for a moment we forget about our present scientific achievements and go back in history to the fifth century B.C., to Athens: one of the greatest centuries in the history of human thought. But first we must set the scene aright. The scene is set in prison. Socrates has been condemned by his fellow citizens on a charge of atheism, impiety, and corrupting the young by his sophistry. In a few hours he is to be given poison to drink. He spends these last hours in discussion with his friends over this very problem of the relation of the mind to the body. He tells them a little of the development of his own thoughts on this subject. Let us listen for a moment to him. Socrates speaks:

GDW

When I was young I had a great desire to know that
department of philosophy which is called 'natural
science'. To understand the causes of things. To
understand why a thing is, how it is created and how
destroyed, this appeared to me a worthy investigation.
I was always preoccupied with such questions as these:
Is it some form of fermentation which causes heat and
cold to bring forth living creatures? Is the blood the
essential element without which thinking could not
occur, or is it respiration, or the natural heat of the
body? Or perhaps none of these but the brain may be
the natural seat of our senses of hearing and sight and
smell; and from these sensations memory and
opinion arise, and then when memory and opinion are
firmly established natural science may be built up.
Then I went on to consider how these powers are lost,
and this led me to consider all phenomena in heaven
or on earth. Finally I came to the conclusion that I
had no aptitude for these studies, and I will tell you
what led me to this conclusion. For I found that my
preoccupation with these investigations had so blinded
my eyes to things which before had seemed to me and
others self-evident.

Doesn't this sound very familiar to you? Pavlov,
Eccles, you and I, starting off with the enthusiastic
conviction that the investigations of natural science
will provide us with answers to the deepest questions.
And then as our investigation proceeded we found
ourselves bewildered and in doubt over what at first
had seemed so certain and well established. This
bewilderment and confusion we found at last arose
because we were trying to solve a philosophical puzzle-
ment by an irrelevant empirical investigation. Socrates
too discovered in good time that these sort of investi-

gations could never answer for him the most important question as to the nature of man and his destiny. I will not here discuss the method of philosophical enquiry that he then developed, but I will proceed to tell you of the conviction he finally came to, and with which he went willingly to his death. Socrates speaks again:

So long as we keep to the body and our soul is contaminated with this imperfection, there is no hope of our attaining our object, which we assert to be absolute truth. For the body is a source of endless trouble to us by reason of the mere requirement of food; and is liable also to diseases which overtake us and impede us in the search after true being: it fills us full of loves and lusts and fears and fancies of all kinds, and endless foolery, and in fact, as men say, takes away from us the power of thinking at all. Whence come wars and factions and fighting? Whence but from the body and the lusts of the body? Wars are occasioned by the love of money, and money has to be acquired for the sake and in the service of the body; and by reason of these impediments we have no time to give to philosophy; and last and worst of all, even if we are at leisure and betake ourselves to some speculation, the body is always breaking in on us causing turmoil and confusion in our enquiries, and so amazing us that we are prevented from seeing the truth. It has been proved to us by experience that if we would have pure knowledge of anything we must be quit of the body, the soul in herself must behold things in themselves; and then we shall attain the wisdom which we desire and of which we say we are lovers. Not while we live but after death. For if while in the body, the soul cannot have pure knowledge, one of two things follows, either knowledge is not to be attained at all, or, if at all, only after death. For then

and not till then, the soul will be parted from the body
and exist in herself alone. In this present life I reckon
that we make the nearest approach to knowledge
when we have the least possible intercourse with the
body and are not surfeited with the bodily nature,
but keep ourselves pure until the hour when God
himself is pleased to release us. And thus having got
rid of the foolishness of the body we shall be pure and
hold converse with the pure, and know of ourselves
the clear light everywhere, which is no other than
the light of truth.

But O, my friend if this be true there is great
reason to hope that, going where I go, when I have
come to the end of my journey, I shall attain that
which has been the pursuit of my life.

For Pavlov 'mind dependent on body': for Socrates
the body a hindrance, a source of distraction and
deceit, an imprisonment for the mind. Socrates of
course, or maybe Plato speaking in the name of
Socrates, is ennunciating a conception which already
had had a long history in Greek thought. There was
indeed a familiar Greek proverb σῶμα σῆμα, 'the body
a tomb'. And it would have been no paradox to his
audience when Euripides put into the mouth of his
chorus these lines:

Who knows if life is not death
and death is considered life in the other world.

What has interested me though, and what I now
want to draw your attention to, is the way this
Pythagorean-Socratic-Platonic conception of the relationship of the soul to the body finds perfect expression
in the most unexpected places. An Elizabethan actor-
dramatist, a man described as knowing little Latin and

less Greek; a man living in an age of intense religious controversy who yet nowhere in his many writings gives us so much as a hint as to where his own allegiance lay. And then suddenly this: *The Merchant of Venice*, Act V, Scene 1. Lorenzo and Jessica have come out into the garden, it is a bright starlit Italian night. Lorenzo speaks:

> Sit, Jessica; look how the floor of heaven
> Is thick inlaid with patines of bright gold:
> There's not the smallest orb which thou behold'st
> But in his motion like an angel sings:
> Still quiring to the young-eyed cherubims;
> Such harmony is in immortal souls;
> But while this muddy vesture of decay
> Doth grossly close us in, we cannot hear it.

Ah! there you see it comes again, 'this muddy vesture of decay doth grossly close us in'. The mind prevented by the body from perceiving the truly real in all its wonder and beauty.

And if anyone should reply that Shakespeare is only putting words into the mouth of one of his many characters, listen to this sonnet of his where he is surely speaking for himself.

> Poor soul, the centre of my sinful earth,
> Fool'd by these rebel powers that thee array,
> Why dost thou pine within and suffer dearth,
> Painting thy outward walls so costly gay?
> Why so large cost, having so short a lease,
> Dost thou upon thy fading mansion spend?
> Shall worms, inheritors of this excess,
> Eat up thy charge? Is this thy body's end?
> Then, soul, live thou upon thy servant's loss,

And let that pine to aggravate thy store;
Buy terms divine in selling hours of dross;
Within be fed, without be rich no more:
So shall thou feed on Death, that feeds on men,
And Death once dead there's no more dying then.

For me though the most impressive and deepest
expression of this Socratic-Platonic conception in all
literature is found in the most unlikely of places.
Victorian England; a dreary Parsonage high up on
the Yorkshire moors; the mother long since dead; the
father a man of no great ability; the brother already
a victim of alcohol; and then three sisters of superlative
imagination. Of Emily Brontë her sister said: 'Stronger
than a man, simpler than a child, her nature stood
alone.' So then hear this, which must surely reflect a
personal experience:

But first a hush of peace, a soundless calm descends;
The struggle of distress and fierce impatience ends;
Mute music soothes my breast, unuttered harmony
That I could never dream till earth was lost to me.

Then dawns the invisible, the unseen its truth reveals;
My outward sense is gone, my inward essence feels:
Its wings are almost free, its home its harbour found,
Measuring the gulf it stoops and dares the final bound.

O dreadful is the check, intense the agony,
When the ear begins to hear and the eye begins to see;
When the pulse begins to throb, the brain to think again;
The soul to feel the flesh, and the flesh to feel the chain!

Yet I would lose no sting, would wish no torture less;
The more that anguish racks, the earlier it will bless;
And robed in fires of hell, or bright with heavenly shine
If it but herald death, the vision is divine.

Poetry! Poetry! What am I doing quoting it here? Am I not speaking to a society dedicated to scientific investigation and the sober weighing of experimental evidence? I tried to demonstrate to you in the first part of my paper that the methods of scientific investigation were not and never could be applicable to this great question, the relation between mind and body. And that which seemed so obvious to Pavlov was far from being self-evident, that indeed when rigorously pursued it led us into obvious nonsense. I am certainly not claiming that the conception which Socrates discussed at his death, and which these poets so unexpectedly echoed, is one that can be proved or verified. The very notion of proof or verification is misunderstanding and superficiality here. I can only say that this 'idea' is like an arrow in the mind. Once it has lodged there it cannot be extracted. That is the reason why poetry is its true formulation. Perhaps this is what Goethe meant when he said that he who does not believe in the world to come is already dead in this one.

But concerning that which can be neither proved nor verified is not a healthy agnosticism the proper attitude in a scientific age? Yes certainly, a healthy agnosticism concerning that which has not yet been ascertained but may perhaps in the future be known, for instance the possibility of life on other planets. But agnosticism has no meaning when applied to those questions which by their very nature will never be a matter of scientific investigation. These are questions, I say, that the conduct of life demands an answer from us now, at once. To suppose that conduct can be divorced from speculation or that we may do good without caring about truth, is a danger that is always tempting.

So now let me draw your attention to certain broad principles in our own field of psychiatry where a decision on this ethical question is urgent and imperative. And yet where neither common sense, nor further information, nor any scientific discovery, can ever come to our aid. Where the will alone must decide the truth it will believe.

Some of us have to take care of and do as much as lies in our power for those who either by genetic defect or birth trauma will never develop into maturity. Some indeed who will never learn to speak or even be able to carry out the simplest acts of self-preservation. Now if Pavlov was correct and mind was dependent on brain, we must assume that where there is such gross brain damage, mind is almost nonexistent too. I have lived through an age in which a great and cultured nation deliberately acted as if this were so; and counted it wise and praiseworthy to destroy such apparent monstrosities. Knowing the history of man I see no reason to be optimistic that our children, or our children s children, may not have to fight this same battle over again. But suppose that what Socrates contended for was indeed the truth. That the soul is *imprisoned* within the body. Then we can say nothing concerning the hidden life of these sufferers; they are shut off from us by barriers that neither we nor they can break; but we do not know yet what they shall be. I state this not as an hypothesis that one day might be proved, nor as anything that some special insight could reveal, but as a decision of the will, a decision of ethics where neither physiology nor any other science can come to our aid.

In the practice of psychiatry we are dealing every day with those whose personality has undergone a change. People who have become morose and depressed; people who have become wildly excited and overactive; people who have become withdrawn and suspicious; people who have become deluded and even dangerous; and so on through all the range of mental illnesses. We have been discovering these last thirty years to what extent these disorders can be cured by purely physical methods of treatment (methods which require, however, patience and explanation in their application). But I think the very success of these methods are to some degree a danger to those who employ them. Whatever advances are made in the future regarding the treatment of mental illness, however close the work of a psychiatrist becomes to that of a general physician, we should never forget that there is, and always will be, a mystery about mental ill-health which makes it different from any disease of the body. Every mentally ill patient is an individual enigma, and we should always think of him as such. There is something more disturbing and puzzling in a dissolution of the personality than in any bodily disease. I think that great and good man Dr Samuel Johnson spoke for all mankind when he described his own experience: 3 a.m. on the morning of 16 June 1783.

> I felt a confusion and indistinctness in my head which lasted I suppose about half a minute. I was alarmed and prayed God that however He would afflict my body He would spare my understanding. This prayer that I might try the integrity of my faculties, I made

in Latin verse. The lines were not very good, but I knew them not very good: I made them easily and concluded myself to be unimpaired in my faculties. Soon after I perceived that I had suffered a paralytic stroke, and that my speech was taken from me. I had no pain and so little dejection in this dreadful state that I wondered at my own apathy, and considered that death itself when it should come would excite less horror than now seems to attend it.

I think Dr Johnson's relief that only his body was affected and not his reason is something that we who have to treat those whose minds *are* affected should constantly remember. For the patient a mental disease is and always will be, whatever advances in treatment are made, a more terrifying and humiliating experience. I think we should make it clear that although we do not share their pessimism about the outcome, we do appreciate their natural alarm.

It was once said to me, 'What I should fear if I became mad would be your common-sense attitude, that you would seem to take it as a matter of course that I was suffering from delusions.' I think I understand what he meant, and I think he was referring to an attitude that it is only too easy for those dealing daily with mental illness to fall into. I believe that we must let our psychiatric patients see that we understand that they are in a state of affliction which is not comparable to any bodily pain however severe. To communicate such an understanding is not easy.

When I was a medical student, the treatment of mental illness was largely a matter of protective custody, attention to physical health, and a patient hopefulness. You younger psychiatrists of today can hardly imagine the mental hospitals of those days. Now on all sides there are treatments to be got on with and you can feel a genuine optimism concerning the ultimate recovery of most of your patients. There is perhaps a danger that we should take all this too much for granted. Many generations of physicians have desired to see the things that you see and have not seen them. I know that with any scientific discovery in time it must lose its wonder; but if what I have been saying concerning the relation of mind and body has carried any conviction to you, then I think these methods of treatment should always be a source of wonder. There is something to wonder at here in the return of sanity; these treatments are on a different plane than any other medical procedure. As time goes on it is probable that we will come to know, and make use of, considerably more about the biochemistry of the central nervous system. The information that the electro-encephalogram can give us is still perhaps only in its infancy. Yet assuming that the steady therapeutic progress of the last thirty years will continue (and remember this is an assumption), there will always be in psychiatry the realm of the inexplicable. An inexplicable which does not exist in any other branch of medicine. There is still, for instance, a great deal to learn about, say, the action of the tri-cyclic drugs on the biochemistry of the brain. But no discovery can ever be made as to how these drugs can relieve melancholia and change nihilistic delusions. This leap from the physical to the mental will remain always in the realm of the inexplicable. Concerning this may I not once again use the word 'the miracle'? I might mention at this point a matter of less importance,

yet which is relevant to what I have been con-
tending for. I am sorry that the word 'psychiatrist' has
come into general use to denote those physicians who
are concerned in the treatment of mental illness. It
suggests I think to the general public, and we may
even deceive ourselves by it, that we have both more
power and understanding than we really possess. None
of us are able to 'heal the soul' as the word psychiatrist implies. I prefer that old-fashioned word 'alienist'.
We are concerned with those who in some way are
alienated from their real selves. We have found in
recent years certain ways of treating the body that
hastens in many cases the return from alienation, but
why this should be so is a matter that will always
defy explanation, just because consciousness and
personality are matters which the notion of 'explanation' is not applicable to.

We have been talking about drugs, their known
action on the human nervous system, and their inexplicable action on the mind and on personality.
There has been talk in recent years of drugs that
might provide new and deeper 'insight' into the real
nature of the world; opening as it were 'the doors of
perception'. There have been those who have advised
and attempted to use such drugs as mescalin and
lysergic acid to obtain a vision of the world freed
from the everyday categories through which we
normally perceive it. In so far as I may have seemed
to speak of and quote those who longed for some such
release –

Its wings are almost free, its home its harbour found,
Measuring the gulf it stoops, and dares the final bound.

– it might seem, I say, as if I would be in favour of such experimentation. I must explain the terrible errors that are present in this way of thinking, which is perhaps also becoming a way of acting. I will leave aside the purely pharmacological aspects of this use of drugs, merely mentioning that at present we have no such drugs whose influence is always beneficial and which carry no risk of addiction. For myself I doubt whether any substance that consistently produced euphoria could be free from the risk of addiction. But suppose in the future a chemical substance was discovered that had all the advantages which Aldous Huxley wrongly claimed for mescalin. I will just remind you of these claims by quoting Huxley's own words:

> These better things may be experienced (as I experienced them) outside or 'in here', or in both worlds, the inner and the outer simultaneously or successively. That they are better seems to be self-evident to all mescalin takers who come to the drug with a sound liver and an untroubled mind.

The fact that Huxley's claim for mescalin is inaccurate is not the most important point here; it is the enormous ethical error that is most in need of exposition. The escape from the body and its limitations that Socrates spoke to his friends about, that Shakespeare and Emily Brontë so impressively expressed in poetry, this in its very essence was something that was given, unearned and unexpected. If it was something that we human beings could manipulate, that was ours to achieve as and when we wanted, then it is not that of which these wrote and of which I was speaking. This is that which must be longed for in expectation and patience. All pleasure-seeking is the search for an artificial paradise, an intoxication, but of this freedom it was truly said 'The wind bloweth where it listeth, and thou canst not tell whence it cometh or whither it goeth.'

You remember the quotation from Socrates' speech finished: 'If no pure knowledge is possible in the company of the body, then it is either totally impossible to acquire knowledge, or it is only possible after death.' The sonnet of Shakespeare and the poem of Emily Brontë both spoke of death with a certain longing and sense of expectation. An entrance into that state for which they longed.

As practitioners of medicine we have to be acquainted with death. It is our duty to fight to the last for the lives of our patients. But we have another duty too, one that is hard to combine with the previous one. And that is to recognise the signature of death when it is inevitable. Did you ever read Macaulay's account of the death of King Charles II? It is a horrid picture. How the Royal physicians clustered round him like flies; they bled him, they repeatedly purged him, and gave him disgusting emetics, until at last the poor king said wanly, 'You must pardon me gentlemen, I seem to be an unconscionable time in dying.' How one would have liked to drive those leeches out of the sick-room and let the poor soul depart in peace.

If what I have said concerning the relation of mind and body is the truth (and remember I have made it my principal endeavour to show that nothing in the nature of proof or reasonableness or evidence has any place here. The will must decide). But if this is your decision, then the moment of death is the supreme moment of life. The moment when the prisoner escapes out of the prison house, as it were a bird out of the snare of the fowler. Then I say we, as physicians, must have insight to know when our work is done to the uttermost. When it is our duty to stand aside and interfere no more.

I would end this paper with one more quotation from Plato; from the dialogue *Phaedrus*. I choose this quotation because it expresses for me so profoundly the mystery of mind and body; the mysteriousness of our present being.[1]

Thus far I have been speaking of the fourth and last kind of madness, which is imputed to him who, when he sees the beauty of earth, is transported with the recollection of true beauty; he would like to fly away, but he cannot; he is like a bird fluttering and looking upward and careless of the world below; and he is therefore thought to be mad. And I have shewn this of all inspirations to be the noblest and highest and the offspring of the highest to him who has or shares in it, and that he who loves the beautiful is called a lover because he partakes of it. For, as has been already said, every soul of man has in the way of nature beheld true being; this was the condition of her passing into the form of man. But all souls do not easily recall the things of the other world; they may have seen them for a short time only, or they may have been unfortunate in their earthly lot, and, having had their hearts turned to unrighteousness through some corrupting influence, they may have lost the memory of the holy things which once they saw. Few

[1] Jowett's translation.

only retain an adequate remembrance of them; and they, when they behold here any image of that other world, are rapt in amazement; but they are ignorant of what this rapture means, because they do not clearly perceive.

CHAPTER 4

HYPOTHESES AND PHILOSOPHY

> DO not call it hypothesis, even less theory, but the manner of presenting it to the mind.
>
> LICHTENBERG

Ladies and gentlemen,

I am going to begin with a quotation from Macaulay's essay on Francis Bacon. It is as follows:

Suppose that Justinian when he closed the schools of Athens, had asked the last few sages who haunted the Portico, and lingered round the ancient plane trees, to show their title to veneration: suppose that he had said: a thousand years have elapsed since in this famous city Socrates posed Protagoras and Hippias; during those thousand years a large proportion of the ablest men of every generation has been employed in constant efforts to bring to perfection the philosophy which you teach; that philosophy has been munificently patronised by the powerful; its professors have been held in the highest esteem by the public; it has drawn to itself all the sap and vigour of the human intellect; and what has it effected? What profitable truth has it taught us that we should not equally have known without it? What has it enabled us to do which we should not have been

HDW

equally able to without it? Such questions we suspect would have puzzled Simplicius and Isidore.

Ask a follower of Bacon what the new philosophy, as it was called in the time of Charles the second, has effected and his answer is ready. It has lengthened life, it has mitigated pain; it has extinguished diseases; it has increased the fertility of the soil; it has given new security to the mariner; it has furnished new arms to the warrior; it has spanned great rivers and estuaries with bridges of form unknown to our fathers; it has guided the thunderbolt innocuously from heaven to earth; it has lighted up the night with the splendour of the day; it has extended the range of human vision; it has multiplied the power of the human muscles; it has annihilated distance; it has facilitated intercourse, correspondence, all friendly offices, all dispatch of business; it has enabled men to descend to the depths of the sea, to soar into the air, to penetrate securely into the noxious recesses of the earth; to traverse the land in cars which whirl along without horses, and the ocean in ships which run against the wind.

You may well be wondering what this long bit of Victorian rhetoric has to do with the title of my paper, Hypotheses and Philosophy. I have chosen it because it illustrates so superbly the central error I want to discuss with you. The confusion as to what the proper function of philosophy is. Why we human beings need it so much, and perhaps particularly in this present age. And why it never hands over a finished result to be transmitted from one generation to another.

My main thesis is this. That a philosophy which takes no cognisance of science becomes empty; and a natural science which is not subjected to philosophical criticism becomes blind. I have chosen the modern mutation-selection theory of evolution to illustrate this thesis. Not that the theory of evolution has any special status in this respect. Modern astronomy when it talks about the ultimate nature of the universe, modern physics when it talks about the fundamental constituents of matter, modern psychology when it talks about the scientific study of personality – any one of these would have served my purpose as well.

But at the present moment the theory of evolution is particularly in need of a little dose of philosophic doubt. It is one of those recent advances in knowledge which appear to be so much more important than they really are. It is a subject in which it is so difficult to say only just as much as we really know.

Perhaps I can make clearer what I want to say if I begin by stating the main line of my argument in rather general terms, and only later fill in the concrete details: as it were formulate the charge against the prisoner at the bar and then proceed to examine the witnesses.

The great philosophical danger in every natural science is to confuse an hypothesis with a fact. A new branch of natural science begins because of new observations, new phenomena not noticed before. Often this is due to the discovery of a new instrument, a telescope, a microscope, an electric cell, a Wilson cloud chamber. But always the new data are perceptions. There is nothing in science which was not first in the senses. Now to communicate these new discoveries and to pass them on to the next generation, a new language is required, new words, new concepts, but most important of all new schemata: models, pictures, maps. These new models, pictures, maps are scientific hypotheses. They are not given to us as necessities, never dictated by the facts, never forced upon us, but invented by us as ingenious abbreviations to summarise the complexities of the mass of new factual data. Which of a large number of possible hypotheses we accept is at first a matter of choice. It is determined to a great extent by the spirit of the age in which the new discoveries are made.

But then when an hypothesis has become generally accepted and shown its usefulness, it forgets its humble origin. It begins to masquerade in the logical status of a fact. Something we can't query. Something which is the reality behind phenomena. Something which has enabled us to see behind the curtain of sensation. And so the hypothesis which is our own useful creation, dazzles our view of things. We fail to see much that the hypothesis doesn't include; we extend the limits of our hypothesis into regions of phantasy. Reality which lies before us at every moment is replaced by the abstract picture we have ourselves created. Reality we are told is nothing but a fortuitous concatenation of atomic particles. Reality we are told is the immense system of extra-galactic nebulae. Reality we are told is that long process of evolution from amoeba to consciousness. In speaking like this we have become dazzled by our picture making.

Now to make this specific charge more precise by considering in detail the modern mutation-selection theory of evolution.

I bow to no one in my admiration for Charles Darwin. Where else will you find such close and accurate observation of plants and insects, of birds and mammals, and the constant interrelation in the lives and deaths of all these creatures? What powers of observation he had!

I bow to no one in my admiration for Gregor Mendel.

Those simple but painstaking experiments with his dwarfed, wrinkled, yellow, tall and short peas. Mendel's demonstration of how already existing characteristics emerged or failed to emerge in the offspring of a particular union; this was indeed a new field of observation. It shows what a real talent for research can do with the simplest of material, and with no financial endowment.

But now on the basis of Darwin's and Mendel's work has grown up what is known as the mutation selection theory of evolution. The theory that the development of all the multitude of living forms both in the vegetable and animal world can be explained in terms of genetic mutation and the survival of the fittest. New forms arise by mutation and survive by natural selection.

For instance in a recent popular book on the evolution of man, I find it stated that 'biologists are no longer interested in finding a proof of this theory, it is now only a matter of filling in the details'.

In more scientific language Professor Medawar states,

It is the great strength of the Darwinian selection theory that it appeals to the working of no mechanisms which are not severally well understood and demonstrable. Selection does occur, that is the members of a population do make unequal contributions to the ancestry of future generations; new variants do arise by the process of mutation.

But Julian Huxley is even more bold; he writes:

> One of the major achievements of modern biology has
> been to show that purpose is apparent only, and that
> adaptation can be accounted for on a scientific basis
> as the automatic result of mutation and selection
> operating over many generations. In Darwin's time
> natural selection was only a theory, now it is a fact.

Thank you, Mr Huxley, for putting so concisely the
logico-philosophical error I wish to refute. A theory
can never become a fact. An hypothesis remains an
hypothesis to all eternity. It always contains an
element of choice, one way of looking at things; one
way of arranging an arbitrary selection of material
into a coherent picture.

The danger of forgetting this is that we proceed to
overlook the facts that *won't* fit into the picture; and
we extend the picture to cover aspects of experience
to which it has *no* relevance. Let me illustrate these
dangers in the case of the mutation-selection theory
of evolution.

First the facts which won't fit into the picture. In
this part of my paper I am largely borrowing from an
important book by Professor C. P. Martin of McGill
University. Time will only allow reference to some of
the salient points in his work.

The two most effective ways of producing mutations
experimentally are the use of X-rays and of nitrogen-
mustard. These are at the same time two of the most
powerful protoplasmic poisons known. All mutations
produced by these agents, that is all experimentally
produced mutations, lower the fertility and viability
of the species so changed. I can find no reference to a

mutation produced by human interference that is not either lethal or sublethal. For instance it is possible to produce by experimental mutation a tailless variety of mouse. But such a breed cannot be continued for more than one or two generations; not that the loss of a tail is so serious a disability, but because the process of mutation has so undermined the viability of the species.

Yet geneticists continue to assert that the millions of variations found in nature arose by mutation; and that these mutations had in certain circumstances increased viability and adaptive value. This is pure guess-work. Fisher, a leading exponent of the mutation-selection theory, really admits as much when he has to write as follows:

> We may reasonably suppose that other less obvious mutations are occurring which at least in certain surroundings or in certain genetic combinations might prove themselves to be beneficial.

Notice those words, 'suppose', 'might', 'at least'. How can Huxley claim that the hypothesis has now become a fact?

Dobzhansky, another protagonist for the theory I am criticising, goes so far as to say:

> The genetic theory of evolution would be embarrassed if anyone were to observe the origin of a mutant superior to the ancestral type in the environment in which the latter normally live.

I therefore assert that what we really know is that mutation is a pathological process, and we are only guessing when we say that it has ever been otherwise.

Now consider the process of natural selection. The

survival of the fittest. Undoubtedly at certain times and under certain circumstances such selection has occurred. And the study of the way in which form and function, structure and coloration, adapt an organism to the complexities of its environment is a fascinating study. But that all the immense variety spread out over the whole face of nature; that all this multi-formity of shape and pattern and habit; that all this is due entirely to a process of natural selection – this seems to me to be the most far-fetched assumption.

Indeed there are many cases where we can see patently that this could not be true. Once again I am borrowing largely from Professor Martin. He devotes a whole chapter to the universal phenomenon of the atrophy of disused organs. Such atrophy proceeds by steps too small for any of them to count as an advantage, and proceeds far beyond the point at which any process of selection could apply.

Herbert Spencer was much intrigued by the size of a whale's femur. Buried deep in the huge carcase of the whale is a tiny bone weighing about two ounces. It is the exact homologue of the femur, the largest bone in the mammalian skeleton. We know nothing about the process by which certain mammals reverted to a purely marine existence. The earliest skeletons of whales are found in the Oligocene formations, and differ little from our present species. Before this period the geological record is completely silent. But suppose they are descended from animals that at one time had legs, and that as the legs became an encumbrance in their new aquatic environment, they gradually atrophied. It could surely make no difference to the survival of the whale if its femur weighed twenty ounces instead of two. The atrophy has proceeded far beyond the point where natural selection could apply.

Exactly the same line of argument applies to the atrophy of the wings of flightless birds. On isolated oceanic islands flightless birds are found which belong unmistakably to species which elsewhere have the power to fly. If a bare incapacity to fly was what natural selection favoured (and it is difficult to see how such an incapacity could be an advantage) natural selection cannot explain an atrophy which has proceeded far *beyond* the capability to fly.

Such considerations lead one on to consider the vexed question of the inheritance of acquired characteristics. The mutation-selection theory seems to consider any such supposition as utterly unscientific. Since the time of Weissman it has been a scientific dogma that all inheritance must be transmitted through the germ cells; and as these are uninfluenced by any experience in the life history of an organism, no acquired characteristic can be inherited. But I can see no reason to believe that all inheritance must be via the germ cells. I see no reason why there should not be psychological factors in inheritance as well as physical. Why certain tendencies, habits, likes and dislikes, should not be directly inherited without being dependent on any material structure for that inheritance. Professor Martin has put this point so well that I would like to quote him. He writes:

> All living creatures form habits. They develop preferences in all their activities and these preferences are transmitted in a measure from generation to generation to generation. In this way biological races are formed. The distinguishing characteristics of these

biological races are not simple modifications, that is individual characteristics, for they do not appear fully in the first generation placed in the environment concerned. They develop progressively in the course of several generations if the race continues to reside in the appropriate environment, and fade out in a similar way if a biological race is transferred to a different environment.

What Professor Martin is here saying seems to me to be of the greatest importance. There is much factual evidence that all living creatures can inherit a psychological aptitude to develop with increasing ease a new habit. What is inherited is the ease in acquiring an acquired characteristic.

Let me give you an example. For about one hundred years the wild Norway rat has been used in laboratory experiments. It has gradually become easier and easier to tame each newborn generation. Richter describes the difference between the wild and laboratory rat in these words:

> The wild rat is fierce and aggressive, attacks at the least provocation, and is highly suspicious of everything in its environment. The domesticated rat is tame and gentle and will not bite unless actually injured.

Tameness is an acquired characteristic. But tameability is inherited. A rigid dichotomy of characteristics into those that are either genetically inherited or acquired during the life-time of the individual is inadequate to describe the facts. It is sheer dogmatism to assert that all inheritance must be transmitted through the genes of the germ cells: that psychological traits must be dependent on anatomical structure. Weissman's theory is nothing but the old fallacy of epiphenomenalism dressed up as a piece of biological science. I would like to say, that the mind has genes of its own that the germ cells know nothing about.

I have been illustrating the danger that arises when a scientific hypothesis takes to itself the airs and graces of a fact. It blinds us from seeing much that won't fit into the hypothesis. But the second danger is more serious.

An hypothesis which is taken for a fact easily assumes an ontological status apart from the data which gave it birth. It becomes a hidden reality behind phenomena. And so we get, in the case of our chosen example, Evolution spelt with a capital E. A recent book by a famous palaeontologist illustrates this confusion very well. I refer to De Chardin's *Phenomenon of Man*. In his Introduction to this book Julian Huxley tells us that De Chardin was delighted with the phrase 'In modern scientific man evolution is at last becoming conscious of itself'. The fundamental idea in the book, if I understand it rightly, is that the long centuries of evolution have at last produced a phenomenon 'consciousness' which is able to understand the process from which it has arisen. Julian Huxley regards this as such a profound conception that he sees in it the foundation of a new humanistic religion. Evolution having become conscious of itself can now plan its own future.

But what a mix up of categories is here. Animal, vegetable, mineral, and – consciousness. Don't you feel there is something wrong about a classification like this? Go back to first principles. Every scientific hypothesis depends on data. And, whatever instruments we use to obtain these data, they are in the end dependent on the use of our senses. There is nothing in science which was not first in the senses. The data of every natural science are data for consciousness. You cannot then bring consciousness in as one of the items of the hypothesis. The material used in the foundation cannot at the same time form the coping stone of the roof. Consciousness is not just one of the things we are conscious of.

Look at it this way. I suppose we have all some time or other been fascinated by looking at one of those pictures of the world as it was, say, in the Carboniferous age, when the coal measures were laid down. Those strange tree-like ferns growing in the swampy deltas of the carboniferous rivers. And as we look at these pictures we almost seem to feel the warmth of those sub-tropical times, and hear the wind rustling that strange foliage, and smell the putrefaction of that marshy land, and to see the play of colours as the sunlight comes streaming down through the matted vegetation. But then at once those old familiar questions come crowding in. Were there any smells there when there was no nose to smell them? And were there any sounds there when there was no ear to hear them? And were there any colours there when there was no eye to see them? And if I now try to take refuge in a theory of primary and secondary qualities, then I am reminded of what I once read in the first chapter of Bradley's *Appearance and Reality*. For this chapter showed me conclusively that such a theory cannot be taken as more than a working hypothesis.

It is doubtless scientific to disregard certain aspects when we work; but to urge that such aspects are not fact, and that what we use without regard to them is an independent real thing – this is barbarous metaphysics.

A pre-historic world which was only a re-arrangement of electrons and protons would be one that we could scarcely attach much meaning to, and it would have certainly lost all its imaginative compulsiveness. You see, what we picture when we construct in imagination the theory of geological evolution, is how the world would have looked to a *mind* capable of being a spectator of all time and all existence. So once again you can't bring consciousness in right at the end and say that it itself is the product of evolution.

We have in immediate experience our one sole contact with reality, and everywhere this immediate experience cries aloud that it is incomplete and fragmentary. Then we go on to construct in imagination the conception of an experience which would be more adequate, more satisfying. That is what every scientific hypothesis, apart from its practical usefulness, attempts to be. And in so far as such a process of inference does bring a greater sense of unity into our experience, it is so far legitimate. What is not legitimate is to think that the process of inference is at an end and the ideal is now reached. In the long run I would say that no purely spatial or temporal picture, no picture of the world consisting of a lot of things scattered about in space and time, can satisfy our demand for a final resting place. For every spatial and temporal picture goes to pieces completely at its edges.

I keep coming back to the fundamental thought of

this paper, the logical status of a scientific hypothesis. That it is always a transitory, incomplete affair. Never finished, final, factual. Every scientific hypothesis is always at the mercy of new evidence and may require indefinite modification in the light of this evidence.

Somewhere between Athens and Marathon there is a great outcrop of Jurassic limestone. The surface of these rocks, so I am told, is studded with fossil shells and bones of the Mesozoic period: the age of the great reptiles. Aristotle must have passed this place many times. Yet I believe I am correct in saying that nowhere in the biological writing of Aristotle is the existence of this place even mentioned. So much the worse for Aristotle you say. All right, so much the worse for Aristotle. But what is sauce for the goose is sauce for the gander. Considering the vast complexity of the matrix of nature, isn't it certain that there is still much evidence, lying before our eyes and beneath our hands, which we have failed to notice as yet? And may not such evidence in the future transform our idea of nature as much as the new biology has transformed the Aristotelian concepts? The great thinkers of the Middle Ages are often criticised in popular works for their subservience to Aristotle. This of course is a gross historical over-simplification. But in so far as it is true, it represents a universal human tendency to take transitory concepts as final and absolute. Huxley and De Chardin, when they make the idea of evolution the basis of their philosophy, even of their religion, are making just the same error. Hypotheses, as Kant said, are contraband in philosophy.

I began this paper with Macaulay's eulogy of natural science. Now at almost the same date that Macaulay was writing this essay a much greater European thinker, Kierkegaard, was writing in his journal this passage:

> There is no use in going in for natural science. There is no more terrible torture for a thinker than to have to live under the strain of having details constantly uncovered, so that it always looks as though the thought is about to appear, the conclusion. If the natural scientist does not feel that torture, he cannot be a thinker, a thinker is as it were in hell until he has found spiritual certainty.

But not only is every scientific hypothesis at the mercy of new data. They all also contain an element of choice. The data we do have can always be interpreted in a number of ways. Consider again the evidence on which the mutation-selection theory of evolution is based: the geological formations and the fossil record. Hegel in his philosophy of nature puts forward the suggestion that the organic forms found in early geological strata never really lived. They are merely anticipations in stone of what was later to be clothed in living flesh and blood. Why do we reject such an hypothesis as foolish and jejune? It is not that we can produce some concrete piece of evidence that refutes it. We do not *know* that a brontosaurus ever breathed or a pterodactyl ever flew. Hegel's hypothesis accounts for all the data. We reject it for two reasons. First because of the way we have been educated. We have been brought up in the Darwinian tradition. Our popular books, our encyclopaedias, our natural history museums, have presented this one hypothesis as a *fait accompli*. It is an hypothesis that is now so familiar that it is mistaken for a fact. But secondly and more important, there is no doubt that the theory of evolution has an immense appeal to our imagination. This appeal is vividly shown by the fact that Tennyson was able to translate these ideas into poetry.

> There rolls the deep where grew the tree.
> Oh earth, what changes hast thou seen!
> There where the long street roars, has been
> The stillness of the central sea.
>
> The hills are shadows, and they flow
> From form to form, and nothing stands;
> They melt like mist, the solid lands,
> Like clouds they shape themselves and go.
>
> .   .   .
>
> 'So careful of the type?' but no.
> From scarped cliff and quarried stone
> She cries, 'A thousand types are gone:
> I care for nothing, all shall go.'
>
> .   .   .
>
> And he, shall he
> Who loved, who suffered countless ills,
> Who battled for the True, the Just,
> Be blown about the desert dust,
> Or sealed within the iron hills?

Spengler claimed that the scientific world view which appeals to the late stages of a culture is closely connected with the architectural forms which inspired the spring time of that culture. The compact, symmetrical, perfectly proportioned Greek temple. And then an historian like Thucydides on the first page of his history saying: 'Before our time nothing much of importance had happened in the world.' So I find myself wondering whether the imaginative appeal, the sense of awe, with which some modern scientific hypotheses fascinate us, those infinite astronomical distances, those long corridors of time peopled with strange monsters, whether this fascination may not be related to the fact that not many generations ago our ancestors found their inspiration in Gothic architecture. Those tall spires reaching into the sky. Those long dark interior perspectives fading into obscurity, where the gargoyles peered out from the stones.

Be that as it may. What I am in earnest about is this. Every scientific hypothesis is a transitory and to some extent arbitrary affair. It must never be allowed to solidify into a pseudo-fact. But why not? What harm is done?

So it is time we got back to Justinian and the question Macaulay puts into his mouth.

'What profitable truth has philosophy taught us that we should not equally have known without it? What has it taught us to do which we could not have equally done without it?'

I would like to think that Isidore replied in the true spirit of Socrates. Good sir, you mistake our purpose. We add nothing to the sum total of human cleverness and skill. Our function is otherwise. When the Delphic oracle told our father founder that he was the wisest man in Athens, he understood this to mean that he alone knew how little he understood. That still remains our function in society. To insist that people say only just as much as they really know; that when, as happens in every generation, new advances in knowledge are made, they are not taken to be more

IDW

important than they really are. You ask what is
the value of such scepticism, such agnosticism, such
carping criticism? One value only. It keeps wonder
secure. That sense of wonder that Samuel Johnson
wrote of in these words:

> We all remember a time when nature gave delight
> which can now be found no longer, when the noise of
> a torrent, the rustle of a wood, had power to fill the
> attention and suspend all perception of the course of
> time.

That sense of wonder that Wordsworth wrote of:

> There was a time when meadow, grove, and stream,
>   The earth and every common sight,
>   To me did seem
>   Apparelled in celestial light,
> The glory and the freshness of a dream.

But remember how Wordsworth ends:

> Turn where so ere I may,
>   By night or day,
> The things which I have seen I now can see no more.

May I end then with a little parable? It is not really
mine; it is taken from one of the novels of Charles
Morgan.

You are sitting in a room and it is dusk. Candles
have been brought in that you may see to get on with
the work in hand. Then you try to look up and out to
the garden which lies beyond; and all you can see is
the reflection of the candles in the window. To see the
garden the candles must be shaded.

Now that is what philosophy does. It prevents us
from being dazzled by what we know. It is a form of
thinking which ends by saying, don't think – look.

CHAPTER 5

MADNESS AND RELIGION

S O what can a man do where he sees so clearly that
what lies before him is not the whole plan?
Answer: No more than work faithfully and actively
on that part of the plan which lies before him.

LICHTENBERG

During the last thirty years a remarkable change has
taken place in the practice of psychiatry. When I was
a medical student there was no known form of treatment for what are called the major psychoses, melancholia, mania, schizophrenia, paranoia. Patients
suffering from these diseases were admitted to hospital
and their physical well-being looked after, but their
recovery from the disease itself was a matter that had
to be left to time and chance. Even in the more fortu-
nate cases this was usually a matter of months or
years. Now for each one of these diseases we have a
specific form of therapy: restraint and seclusion are
things of the past, and duration of stay in hospital is
measured in weeks rather than in months and years.

To most people's surprise these treatments have
turned out to be physical and chemical in nature, and
have not arisen from any deeper understanding of the

psychological processes causing the symptoms mani-
fested. The fact that a person's mood and content of
thought can be so profoundly and rapidly altered by
the administration of a few pills or injections, or by an
artificially induced convulsion, seems to me to raise
important questions both in philosophy and ethics.
These questions I do not find anywhere adequately
discussed. For two reasons I think. Those who are
using these treatments are so delighted to be able at
last to do something positive and effective for their
patients, that they have neither the time nor the
training nor the inclination to raise questions about
first principles and ultimate objectives. Whereas those
who are trained to think dialectically have little oppor-
tunity to see the dramatic way these treatments work.

I am therefore very grateful to you for giving me
the chance to discuss my problems with you. I think
the best way to begin would be to tell you in some
detail about four case histories which I have taken
from the records of the hospital where I work. I would
emphasise that there is nothing very unusual about
the first three cases. I am sure any busy mental
hospital could produce similar ones. The fourth case
is unusual but I have included it as it brings out one
aspect of my problem very clearly.

The first case I want to describe to you is that of a
man aged fifty-four, a priest. We will call him Father
A. This priest had for some years been directed by his
Superior to conduct retreats, a type of work for which
he was considered to have great gifts. A few months
prior to my seeing him he had begun to feel very
depressed about his work, that he could no longer
put any feeling into what he was preaching; that he

was asking people to believe and do things which he himself had lost faith in. It was a great burden for him to say Mass or read his daily office. He felt that he ought never to have been ordained, that he had no vocation. When he visited his brother, a happily married man surrounded by his family, he felt *that* was the sort of life he was meant for. In addition he began to lose weight and to have very disturbed sleep, he would wake about three in the morning and lie awake till dawn worrying about his spiritual state. He developed a feeling of great tension and discomfort in the pit of his stomach. He could not eat. These symptoms led him to believe that he had cancer, to hope indeed that he had cancer and that he would soon be dead. He consulted a physician who advised admission to a general hospital for investigation. After the usual X-rays and biochemical tests had been done he was told that there was no evidence of any organic disease. But he felt no better for this information. It was at this stage that a psychiatrist was called in, who diagnosed an involutional depression, and recommended admission to a mental hospital for treatment. So it was he came under my care. When I first saw him he was resentful and suspicious. His condition was a spiritual one, he stated, and no doctor could aid him. He had brought it on himself and must bear the blame for it. I concentrated on his insomnia and his abdominal pain and asked him to let me treat these symptoms, leaving the whole question of his spiritual state in abeyance for the time being.

I gave him a course of what is known as electric convulsive therapy. It consists in giving the patient an anaesthetic and then passing a current of 150 volts for

about one second through the frontal lobes of the
brain. This causes a generalised epileptic-like convulsion which lasts about two minutes. Within fifteen
minutes the patient is awake and fully conscious
again.

After the first treatment the pain in the abdomen
had gone. He began to eat better, he needed less drugs
to obtain a full night's sleep. Within a week he came
spontaneously to ask if he could say Mass again. By
the time he had seven such treatments he stated he
was feeling very well. He was sleeping soundly without
any drugs and had gained ten pounds in weight. But
this is what is significant: his spiritual problem had
disappeared too. He was saying Mass every morning,
and could read his daily office again with devotion.
He felt ready to return to his work and to conduct
retreats as before. This is what he is now doing, though
his Superior has been advised to see that he has
proper intervals of rest.

A straightforward case of involutional melancholia
properly treated, most of my colleagues would say.
Why do I say that it raises important philosophical
and ethical questions? Well, now listen to this piece
of autobiography, written nearly a hundred years
ago, by a man about the same age as Father A:

I felt that something had broken within me on which
my life had always rested, that I had nothing left to
hold on to, that morally my life had stopped. An
invincible force impelled me to get rid of my existence,
in one way or another. It cannot be said that I wished
to kill myself, for the force which drew me away from
life was fuller, more powerful, more general than any
mere desire. It was a force like my old aspiration to

live, only it impelled me in the opposite direction. It
was an aspiration of my whole being to get out of life.

Behold me then a happy man in good health, hiding
the rope in order not to hang myself to the rafters of
the room where every night I went to sleep alone;
behold me no longer going shooting, lest I yield to the
too easy temptation of putting an end to myself with
my gun.

I did not know what I wanted. I was afraid of life;
I was driven to leave it; and in spite of that I still
hoped for something from it.

All this took place at a time when so far as all my
outward circumstances went I ought to have been
completely happy. I had a good wife who loved me
and whom I loved, good children and a large property
which was increasing with no pains taken on my
part.

I am sure that you feel with me the similarity
between this man's state of mind and that of Father A.
And having seen several hundred such cases recover
with the same treatment that I gave Father A, I cannot help concluding that had such treatment been
available in those days this man's two years of suffering could have been terminated in as many weeks.

But would it have been right to do so?

For the writer of that piece of autobiography was
Count Leo Tolstoy. It occurs in a book which he calls
My Confession. The thoughts and convictions which
eventually delivered him from this misery were to
determine his whole future manner of life and writing.
He says expressly that he was in good health, and I
am sure that like Father A he would have resented
any interference by a doctor.

Again going back a little further in history; when I

read some of the great spiritual directors of the seventeenth and eighteenth centuries, such writers as
Fenelon and De Caussade, they seem to me to be
writing sometimes to people in just such a state of
mind as Father A or Tolstoy. They speak about states
of aridity and dryness, of loss of the faith. Here for
instance is Father Gratry describing his own experience of such a state:

> But what was perhaps more dreadful was that every
> idea of heaven was taken away from me. I could no
> longer conceive of anything of that sort. Heaven did
> not seem to me to be worth going to. It was like a
> vacuum, a mythological elysium, an abode of shadows
> less real than the earth. I could conceive no joy or
> pleasure in inhabiting it. Happiness, joy, love, light,
> affection, all these words were now devoid of sense.

Yet these spiritual directors universally teach that
such states, these dark nights of the soul, are necessary
stages in the growth of spiritual maturity. They are
sent by God, and are to be accepted willingly and
patiently; they are a proof that the soul has now
passed the beginner's stage of sensible consolation,
and is being educated by suffering.

But today it would seem a psychiatrist can treat
such states of mind not out of the abundance of his
spiritual wisdom and experience, but by mechanical
and materialistic means: electrical stimuli to the
brain, drugs which alter the biochemistry of the
nervous system. Such treatments can be given by some
recently qualified young man to whom the spiritual
agony of the patient is something quite outside his
comprehension.

That is why I say such a case as that of Father A raises for me philosophical and ethical problems. Can we differentiate between madness and religion? Can we say of one such state: 'This is a mental illness and is the province of the psychiatrist'? And of another: 'This is a spiritual experience sent by God for the advancement of the soul and is the province of a wise director'?

My second case is that of Miss B, forty-three years of age, living in the west of Ireland and employed as the housekeeper by the parish priest. She was admitted to hospital in a state of elation and excitement. She had had a personal revelation from 'the little flower', St Thérèse of Lisieux. This had occurred when she was visiting a holy well near her home. She had seen lights in the sky which conveyed a special message to her. She had been ordered to convert all the protestants in Ireland. In the ward she rushed across to preach to two non-catholic patients who were there. Some of the junior nurses wore a pink uniform. This was a sure proof that they were half-communists, and she would receive neither food nor medicine from their hands. She denied emphatically that her experiences were in any way due to an illness and resented being in a mental hospital.

Her treatment consisted in a short course of electric convulsive therapy followed by the administration of large doses of a comparatively new chemical substance which has been found to control rapidly such states of exaltation. In three weeks' time her behaviour and conversation were completely normal. She never referred spontaneously to her experiences and only seemed embarrassed when they were mentioned. But she was never willing to admit that it had all been

a matter of illness. She was able to go into the city
alone and always returned to hospital as requested.
And now at the present moment I hope she is cooking
Father Murphy's supper among the quiet hills of
County Mayo.

Nowadays, seeing this patient's state of mind, few
people would hesitate to describe her as mentally ill.
Certainly her parish priest who brought her to hospital
had no doubts on the matter. But in previous ages and
among simpler folk might she not have been regarded
as indeed the recipient of a divine revelation? We were
inclined to smile at her delusion about the nurses' pink
uniforms, but now listen to this:

> I was commanded by the Lord of a sudden to untie
> my shoes and put them off. I stood still for it was
> winter, but the word of the Lord was like a fire in me
> so I put off my shoes, and was commanded to give
> them to some shepherds who were nearby. The poor
> shepherds trembled and were astonished. Then I
> walked about a mile till I came into the town, and as
> soon as I was got within in the town, the word of the
> Lord came to me again to cry 'Woe to the bloody city
> of Lichfield'. So I went up and down the streets crying
> with a loud voice, 'Woe to the bloody city of Lich-
> field'. And no man laid hands on me; but as I was
> thus crying through the streets there seemed to me a
> channel of blood flowing down the streets and the
> market place appeared to me like a pool of blood.
>
> And so at last some friends and friendly people came
> to me and said, 'Alack, George, where are thy shoes?'
> I told them it was no matter.

That was George Fox, the founder of the Society of
Friends. Madness or religion?

My third case is this. Guard C was twenty-seven years of age, a policeman on motor-cycle patrol in the city of Dublin. One day his sergeant was horrified to see Guard C's motor cycle propped up against some railings and the guard himself kneeling in prayer on the pavement. He was taken to hospital in a car and on arrival there was at first quite mute. He appeared to be listening intently to something coming from one corner of the ceiling. His lips moved silently as if in prayer. Later in the ward he stated that a voice from heaven had told him that he had been chosen by God to drive the English soldiers out of the Province of Ulster. He was to be made a commissioner in the Guards and after his death he would be canonised as a saint.

Once again the treatment of this patient consisted in the administration of a powerful chemical substance both by mouth and by injection. Within six weeks he was able to admit that his ideas had been delusions due to illness, and after a proper period of convalescence he was able to return to duty.

But in 1429 when Joan of Arc came to Vancouleurs she stated that the voices of St Michael and St Catherine had ordered her to drive the English soldiery from the fair Kingdom of France. Robert de Baudricourt gave her a horse and a suit of armour; and then – but we all know what happened then. My question is this. Supposing Robert de Baudricourt had been able to give Joan a stiff dose of phenothiazine instead of the panoply of a knight at arms, would she have returned in peace to the sheep herding at Domremy?

The fourth and final case history is this. It came to

my notice over fifteen years ago, when many of the methods of treatment we now have were not known. Mr D was sixty-seven years of age, a retired civil servant, a man of great piety who devoted his retirement to prayer and works of charity. His wife had no sympathy for what she regarded as a morbid religiousness. One morning at Mass he heard read the words of the Gospel: 'Go and sell all that thou hast and give to the poor and thou shalt have treasure in heaven, and come and follow me.' These words spoke to him like a command. And straightaway he left the church, putting all the money that was on him into the poorbox at the door. He set off to walk the 135 miles to Lough Derg, a famous place of pilgrimage in Ireland since earliest times.

When he did not return for his breakfast and the morning passed without news of him, his wife became alarmed and notified the Guards. Eventually that evening he was stopped by a policeman in a small village about thirty miles from Dublin. He was seen by a doctor and put on a Temporary Certificate for admission to a mental hospital. He made no protest at entering hospital, told his story clearly, and accepted what had happened as God's will. I gave this man no treatment other than insisting that he had his breakfast in bed and allowed us to restore a rather emaciated frame. I learnt more from talking to him than he did listening to me. There was at first some difficulty in getting his wife to take him home. She was convinced that he suffered from a condition she called religious mania, but eventually after some weeks she agreed to his discharge.

But now go back a little over sixteen hundred years

and to a church in Alexandria. Another man hears these same words read from the altar. And straightaway he goes out into the desert around Thebes and lives there until his death a life of heroic austerity. Soon thousands are to follow him; to form themselves into communities, to draw up a rule of life. It is the beginning of Christian Monasticism with all that it was to mean for European religion and culture. And so Anthony was canonised and Mr D was certified.

Madness or Religion?

But why should I be quoting from Fénelon and George Fox, from St Joan and St Anthony? I suppose no psychiatrist can read the Bible without sometimes hearing a disturbing echo of what he has just heard said to him on his ward round.

Behold I was shapen in wickedness and in sin did my mother conceive me.

Was this written by someone in a state of melancholia, and would a course of electroplexy have given him a more sanguine estimate of man's estate?

Thou makest my feet like hart's feet and settest me up on high. He teacheth my hands to fight and mine arms shall break even a bow of steel. I will follow upon mine enemies and overtake them, neither will I turn again until I have destroyed them.

Was this written in a state of manic elation and was a sedative called for here?

Thou art about my path and about my bed and spiest out all my ways. For lo there is not a word in my mouth but Thou knowest it altogether.

Schizophrenics often complain that all their inmost

thoughts are being read and controlled by some power outside themselves.

The prophet Ezekiel, the most ecstatic and visionary of the prophetic writers, gives an account of a catatonic state with functional aphonia such as could be duplicated by reference to any standard textbook of psychiatry:

> But thou, son of man, lay thyself on thy left side and I shall lay the guilt of the house of Israel upon thee; the number of days that thou shalt lay upon it shalt thou bear their guilt. And behold I shall lay cords upon thee that thou shalt be unable to turn from one side to the other, till thou hast ended the days of thy boundness.

And in three other passages occur the words:

> In that day shall thy mouth be opened and thou shalt speak and be no more dumb.

In the New Testament too the same problem is thrust upon us:

> And I heard behind me a great voice as of a trumpet speaking, saying: what thou hearest write in a book.

Was the author of Revelations hallucinated? 'Paul, thou art mad,' said Festus, 'thy great learning hast made thee mad.' And did not the Pharisees, those religious experts, say of our Lord, 'Say we not well that thou art a Samaritan and hast a devil?'

Then there is that strange account in St Mark's Gospel, which the other Evangelists omit:

> And when his friends heard of it they sought to lay hands on him, for they said he is beside himself.

Most commentators agree that the friends mentioned here refer to his mother and his brethren who had been mentioned in the previous verse. So you see this problem of ours is one that can deceive even the very elect.

I am sure that by now all sorts of possible answers to the problem have been coursing through your minds. Let us look at some of these answers and see if they will do.

One answer cuts the knot straight away. For Freud there is no problem here. The distinction between the pathological and religious state of mind cannot be made because it does not exist. In his books *Totem and Taboo, The Future of an Illusion,* and *Moses and Monotheism,* Freud argues that it is obvious to anyone trained in psycho-analysis, that religious beliefs and practices are a racial neurosis. The conviction with which such beliefs are held without scientific evidence for them, is the same conviction with which a paranoic clings to his systematised delusion in spite of any proof. The strictness with which religious ceremonies are observed, is the same as that with which an obsessional carries out his profitless repetitions.

I find this simple solution of Freud's entirely unacceptable. Freud never comes to grips with the central problem of ethics. It is clear from reading his biography and personal letters that the man himself was more than his theory. He had a strong sense of duty, and a system of absolute values about which he was not prepared to compromise. A passion to find out the truth, a courage to stand against unpopularity and hostility, a love of nature and art, a lifelong devotion to his wife and children. There is an amusing but I think significant story which Ernest Jones records about Freud.

At the time when the relations between Freud and Jung were almost at breaking point, Jung was still secretary of the psycho-analytical association. He sent Jones an announcement of the next meeting but made an error in the date, so that if Jones had not had other information he would have missed the meeting entirely. Jones knowing Freud's interest in these slips of the pen and tongue showed the letter to him. But Freud was neither interested nor amused. No gentleman, he said, *ought* to have an unconscious like that.

Ought? Ought? Ought? What is that ought doing there on the lips of a psycho-analyst? Of course, we like Freud all the better for this human touch. You see, however much we may exclude oughtness from our theories, we cannot get it out of our lives. Oughtness is as much an original datum of consciousness as the starry vault above. Both should continue to fill us with constant amazement.

Freud's solution of our central problem is, for me, altogether too one-sided to satisfy. It omits entirely to take into account an essential aspect of life – our sense of duty – an aspect which is the very source of the problem itself. When is it *right* to treat this man as mad and when to say let be, let his spiritual growth proceed without meddlesome interference?

I mentioned the Swiss psychiatrist, Carl Gustav Jung, a moment ago. After his break with Freud, Jung developed a doctrine almost diametrically opposed to that of his former teacher. This is what he says in one of his later books: 'Among my patients in the second half of life there has not been one whose problem in the last resort was not that of finding a religious outlook.'

So once again the problem which I raised as to the criterion between madness and religion does not arise for Jung. Madness is religion which has not yet come to an understanding of itself. Madness is the protest of those unconscious needs and forces which the patient has not allowed to find any expression in his life. These unconscious needs are not sexual as Freud taught but religious: all that side of human nature which in the past has found expression in myths and cults and symbols.

I must say that as a theoretical solution this appeals to me. My difficulty arises when I try to make use of it in practice. I would like to be able to cure my patients by discussion, advice, wise counsel, and from an understanding of their spiritual needs. But my experience has been that in all the serious disturbances of the mind such as find their way into a mental hospital, the word has lost its power. Take those first three cases I described to you. I do not know how anyone could talk Father A out of his depression; could convince Miss B that her vision was an hallucination; demonstrate to Guard C that his sense of mission was a delusion. Whereas I do know that by means of these physical methods of treatment I can at least restore them to their former equanimity and return them to their gainful occupations. It is precisely the limitations of these methods that I am debating with you. When to say, 'This man is mad and we must put a stop to his raving,' and when to say, 'Touch not mine anointed and do my prophet no harm.'

One solution which for some time seemed to me to

offer at least a practical solution of our problem was this. You remember that in the case of Father A there was a disturbance of his physical health. Pain, loss of appetite, loss of weight and insomnia. It was these symptoms that enabled me to get his consent to treatment. May we not say then that where there is an obvious failure of physical well-being then we may diagnose morbidity and not spirituality?

But I now find that I must reject this source of distinction. For these bodily disturbances, though common, are not a constant feature in all mental illness. Besides, every psychiatrist knows that these are secondary phenomena. The crux of the matter is the emotional disturbance which has caused them. It is to this that the treatment is directed. Remove the depression, subdue the excitement, get rid of the hallucinations, and sleep and appetite and physical health are restored too.

And then looking at this matter from the other side; the lives of the saints are not free from just these same disturbances of physical health. Von Hügel in his great two-volume study of St Catherine of Genoa has to devote a whole chapter to what he calls her psychophysical peculiarities. This is what he has to say:

Now as to those temperamental and neural matters to which this chapter shall be devoted, the reader will no doubt have discovered long ago that it is precisely here that not a little of the 'Life and Teaching' is faded and withered beyond recall, or has even become positively repulsive to us. The constant assumption and frequent explicit assertion on the part of nearly all the contributors, upon the immediate and separate significance, indeed the directly miraculous character of certain psycho-physical states; states which taken thus separately would now be inevitably classed as most explicable neural abnormalities. Thus when we read the views of nearly all her educated attendants that 'her state was clearly understood to be supernatural when in so short a time a great change was seen and she became yellow all over, a manifest proof that her humanity was being entirely consumed in the fire of divine love', we are necessarily disgusted.

I have quoted Von Hügel at some length to emphasise that someone writing not from a medical but from a theological standpoint, finds this same problem requiring attention.

The reason why we, looking back on history, are able to make the distinctions we do is because of what was achieved. After his conversion Tolstoy, both by his manner of life and his writings, exercised a profound influence on all Europe. George Fox was the founder of the Society of Friends and his influence is with us to this day. And Joan of Arc went forward with the royal banners to the crowning of her king at Rheims. So is it not true that by their fruits ye shall know them, and that it is in this that the distinction we have been looking for will be found?

But we are surely treading on dangerous ground if we introduce results and success into the religious category, and make this our absolute criterion. What sort of results? What sort of success? Is failure and defeat always to be a condemnation? Let me put this matter quite concretely with a particular example.

That great mathematical genius, Blaise Pascal, would almost certainly have preceded Newton and

Leibniz in the discovery of the infinitesimal calculus,
if it had not been for what happened on the night of
Monday, 23 November 1654. You remember his own
description:

> From about half past ten to half after midnight,
> FIRE
> God of Abraham, God of Isaac, God of Jacob, not of
> the philosophers and wise. Security, Security. Feel-
> ing, Joy, Peace. Forgetfulness of the world and of
> all save God. O righteous Father, the world has not
> known Thee, but I have known Thee.

And so Pascal turned from his mathematical studies
to write his defence of the Gospel. These fragments we
now have in his *Pensées*. Those to whom the *Pensées* are
a source of depth and wonder will see in them a proof
of the authenticity of that night of pentecostal fire.

But then I take up a recent history of mathematics
in which the author bewails what he calls Pascal's
nervous breakdown on that fatal night – leading him
to forsake his true genius for what this writer calls
meaningless mysticism and platitudinous observations.
So an attempt to find the distinction we have been
looking for in the results achieved, once again fails us.
For who is to be the judge of the results?

When in philosophy you keep coming up against a
dead end, such as we have so far, in our search for a
principle of differentiation between madness and
religion, it is often because we are looking for the
wrong type of answer. And this indeed is what I
believe we have been doing in our search. For we were
sitting back in a cool hour and attempting to solve
this problem as a pure piece of theory. To be the

detached, wise, external critic. We did not see ourselves and our own manner of life as intimately
involved in the settlement of this question. Now there
can indeed be experts and critics in all the arts and
sciences, but their writ does not run in the realm of the
religious. It is not possible to adopt a detached and
purely theoretical attitude in these matters. It is not
given to any man to be an honorary member of all
religions.

I suppose most thoughtful people have realised that
sooner or later they may be afflicted with some painful, perhaps mortal, disease, and have considered how
they hope to comport themselves under this trial.
But to face clearly the possibility of a mental illness –
that is both too terrible and too much an unknown to
accept. That we, you and I, might one day have to be
admitted to a mental hospital in a state of despairing
melancholia, or foolish mania, might become deluded
or hallucinated – that is a thought not easily to be
entertained. We like to think that either our intelligence, or our will-power, or our piety, would save us
from such a fate. I heard a sermon some time ago in
which the preacher stated that the great increase in
mental illness at the present time was due to the
decay of faith. Anyone, he said, who had a firm belief
in God would never suffer from nerves or a mental
breakdown. Alas his premises were false and his conclusion erroneous. For just those qualities of personality in which we trust, which we regard as peculiarly
our own for keeps, our intelligence, our will-power,
our piety, these are all dependent on the proper
functioning of a very complicated and delicate neuro-
humoral mechanism over which we have no control.

Some slight disturbance of an endocrine secretion, a hardening of some arterial wall, a failure of an enzyme to catalyse an essential chemical reaction, and all in which we have put our trust is gone. Our sanity is at the mercy of a molecule.

So now if we really face these facts squarely the whole problem we have been debating changes, we find *ourselves* closely involved. The problem now is not how as external observers we are to distinguish between madness and religion, but how are we to reconcile the existence of madness, and the ever present threat of madness, with our religious convictions and beliefs.

And then the answer is not far to seek. For it has always been a central doctrine in Christian ethics that the greatest danger to man as a spiritual being does not come from the animal side of his nature, his lusts and passions or even their perversion, but comes precisely from those qualities which distinguish him from the brute creation, his intelligence and efficiency. Pride, self-sufficiency, smugness, 'Lord I thank Thee that I am not as other men are', it is these sins that are utterly stultifying and soul destroying.

Towards the end of his life Kierkegaard wrote in his Journal these words:

> Sometimes in moments of despondency it strikes me that Christ was not tried in the suffering of illness, at least not in the most painful of all where the psychological and the physical touch each other dialectically, and consequently as though his life was easier in that respect. But then I say to myself: Do you think if you were perfectly healthy, you would

easily or more easily become perfect? On the contrary you would give in all the more easily to your passions, to pride if no other, to an enormously heightened self sufficiency.

To lead a really spiritual life while physically and psychologically healthy is altogether impossible. One's sense of well being runs away with one. If one suffers every day, if one is so frail that the thought of death is quite naturally and immediately to hand, then it is just possible to succeed a little; to be conscious that one needs God. Good health, an immediate sense of well being, is a far greater danger than riches, power, and position.

Simone Weil had the same thought too. She writes:

To acknowledge the reality of affliction means saying to oneself: I may lose at any moment through the play of circumstance over which I have no control anything whatsoever I possess, including those things which are so intimately mine that I consider them as being myself. There is nothing that I might not lose. It could happen at any moment that what I am might be abolished and be replaced by anything whatsoever of the filthiest and most contemptible sort. To be aware of this in the depths of one's soul is to experience non-being. It is the state of extreme and total humiliation which is also the condition for passing over into truth.

It is a common prejudice, and one hard to get free from, that a mental illness is a degradation of the total personality; that it renders the sufferer to some degree subhuman. Thus many people would feel that if Tolstoy really suffered from melancholia his challenge to our whole western way of life would be largely blunted and nullified. And if Joan of Arc was a schizophrenic she could not at the same time be a saint. But these are prejudices. A mental illness may indeed utterly disable the patient for the daily commerce of social life, but the terrifying loneliness of such an experience may make him more aware of the mysteriousness of our present being.

A short time ago I was called to see a patient who had just been admitted to hospital for the fifth time. When I came to her she was sitting up in bed reading her Bible with the tears streaming down her face. I thought to myself, this woman understands that book better than I do, or indeed many a learned theologian. 'They that are whole need not a physician but they that are sick.'

Many years ago Ludwig Wittgenstein asked me if I could arrange for him to have conversations with some mental patients. Of one of them, a certified and chronic inmate of the institution, he observed, 'I find this man much more intelligent than any of his doctors.'

There was an old pagan saying, 'Quem deus vult perdere prius dementat.' Perhaps we should baptise that saying. 'Sometimes those whom God intends to save he first has to make mad.'

Every death-bed can be a religious experience both for him who is dying and for those who had loved him and watch beside him. Every mental illness can be a religious experience both for him who is afflicted and for those that loved him. Conversely every religious belief and practice where it is deep and sincere is madness to those who trust in themselves and despise others. That distinction we spent so much time looking for was nothing but a will-o'-the-wisp.

But then what about those mechanical methods of treatment I mentioned at the commencement? Are they not sometimes at least a gross interference with what should be left to the wisdom of God? Are we always right to use them?

Of course we are. A doctor who tries to prolong life and ease the pains of the dying in no way detracts from the majesty and significance of death. A doctor who attempts to shorten and relieve the suffering of the mentally ill in no way diminishes the lesson of madness. If we are to take the doctrine of the creed seriously, 'by whom all things were made', then we must accept that madness in all its horror is as much part of God's creation as the tubercle *bacillus* and the cancer cell. We do not know why these things should be, and if we did they would not be what they are. We are right to fight against them with all the energy and all the weapons that we have. For this energy and these weapons are also part of His creation. But this we must never forget, good physical health, good mental health are not the absolute good for man. These can be lost and yet nothing be lost. The absolute good, the goal and final end of our being is in heaven and not here; and all earthly things as though they get us but thither.

And so to all of us, in sickness or in health, in sanity or in madness, in the vigour of youth or in the decrepitude of senility, God speaks these words which He spoke once to St Augustine:

Currite, ego feram, et ego perducam, et ibi ego feram.

Run on, I will carry you, I will bring you to the end of your journey and there also will I carry you.

INDEX